THE WOMAN'S GUIDE
TO NAVIGATING THE PH.D.
IN ENGINEERING & SCIENCE

Books of Related Interest from the IEEE Press

THE WOMAN'S GUIDE TO NAVIGATING THE PH.D. IN ENGINEERING & SCIENCE

Barbara B. Lazarus
Associate Provost for Academic Affairs
Adjunct Professor of Educational Anthropology
Carnegie Mellon University
Pittsburgh, PA

Lisa M. Ritter
Communications Consultant
Carnegie Mellon University
Pittsburgh, PA

Susan A. Ambrose
Associate Provost for Educational Development
Director of the Eberly Center for Teaching Excellence
Principal Lecturer, Department of History
Carnegie Mellon University
Pittsburgh, PA

IEEE
PRESS

The Institute of Electrical and Electronics Engineers, Inc., New York

This book and other books may be purchased at a discount
from the publisher when ordered in bulk quantities. Contact:

IEEE Press Marketing
Attn: Special Sales
445 Hoes Lane, P.O. Box 1331
Piscataway, NJ 08855-1331
Fax: +1 732 981 9334

For more information about IEEE Press products, visit the
IEEE Online Catalog & Store: http://www.ieee.org/store.

ISBN 0-7803-6037-0
IEEE Order No. PP5883

Library of Congress Cataloging-in-Publication Data

Lazarus, Barbara B.
    The woman's guide to navigating the Ph.D in engineering & science / Barbara
B. Lazarus, Lisa M. Ritter, Susan A. Ambrose.
        p. cm.
    Includes bibliographical references and index.
    ISBN 0-7803-6037-0
        1. Women engineers—United States.   2. Women scientists—United States.
    3. Engineering—Study and teaching (Higher)—United States.   4. Science—
    Study and teaching (Higher)—United States.   I. Ritter, Lisa M., 1962-
    II. Ambrose, Susan A., 1958-   III. Title.

TA157.L385 2000
620'.0071'173—dc21

                                                            00-050600

*To the women engineers and scientists
who mapped the way and to those women
who are navigating their careers to reach a destination
where they, too, belong.*

CONTENTS

"What advice can I give young women today?
I have no mysterious secrets to impart. The most important
advice is to choose the field that makes you the happiest.
There is nothing better than loving your work.
Second, set a goal for yourself.
Even it if is an 'impossible dream' each step toward it gives a
feeling of accomplishment. Finally . . . be persistent, don't let
yourself be discouraged by others, and believe in yourself."

Gertrude Elion
Nobel Prize-winning chemist

*From the keynote address for "Choices and Successes: Women in
Science and Engineering," sponsored by the New York Academy of
Sciences and held March 12–13, 1998, in New York City.*

FOREWORD

Denice D. Denton, Dean of Engineering, University of Washington. (Photo credit: Mary Levin, University Photography, University of Washington.)

If you are reading this book, then you are in the midst of one of the most important times in your professional life. As a graduate student, you can focus on the research and education issues that you care about the most without all of the distractions that you will face later in your career (e.g., endless monotonous meetings, committee work, administrative burdens, management . . .). This is the time for you to really hone in on what you care about and to push the envelope of knowledge in that field or discipline. How many people get that kind of opportunity? Very few.

You have chosen one of the most exciting times in history to study a technical field at the graduate level. There is an emerging notion that the engineering degree is the liberal arts degree of the twenty-first century. Historically, the liberal arts degree was believed to prepare a young person for a very broad variety of career paths. In this century, because of the pervasiveness of high technology and the integration of that technology into the workplace, engineering and physical sciences graduates will be optimally positioned to take on a wide variety of roles. More importantly, you will be better positioned to take on leadership roles in both corporate and academic settings. You should take full advantage of this *carte blanche* as you make career decisions.

In addition to the centrality of science and technology in the global economy, greater numbers of women and people of color are pursuing careers in technical fields. On average, women comprise 20% of the enrollment in engineering schools. At the end of the "educational pipeline," 1,600 women a year receive Ph.D.s in the physical sciences. This represents 24% of such degrees granted. While 769 women received Ph.D.s in engineering in 1998, this represents only 13% of the 5,991 degrees granted in the field. Women make up 56.8% of the U.S. workforce, but only 8.5% of its engineers are women. Despite the increasing representation of women and people of color, the numbers are still small. This may mean that you will be "the first." You might be the first woman in your department to get a Ph.D. in a particular discipline or the first woman in your unit when you choose your first job after graduate school. I was the first woman dean of engineering at a research university (and that was just in 1996!). The great news is that there are now four women deans of engineering at similar universities.

Graduate school presents a number of challenges and potential obstacles. Some of these challenges are a result of the underrepresentation of women in engineering and the physical sciences. This book lays out a "map" of the graduate school experience that will be invaluable to *all* graduate students and mentors in engineering and science disciplines. Also, it specifically addresses successful strategies for women. The book illuminates many important topics, including choosing an advisor, survival skills in graduate school, and planning and preparing for your first professional job after graduate school.

In the end, it's most important that you have fun in graduate school. Try to give talks at international meetings and see other parts of the world. You should enjoy much of the graduate school experience; you can make it what you want it to be. Many jobs you will have later will not have the degree of flexibility and autonomy that you can experience in graduate school. If you take responsibility for your graduate career and make your choices carefully using the advice in this book, graduate school will be one of the most enjoyable and interesting parts of your professional career. Enjoy!

Denice D. Denton
Dean of Engineering, University of Washington
(Ph.D., Electrical Engineering,
Massachusetts Institute of Technology)

A NOTE FROM
THE AUTHORS

The idea for this book came from Barbara Lazarus and Susan Ambrose, administrators/professors at Carnegie Mellon University, who realized that there was no such guide for women working on their doctorates in engineering or science. Barbara, associate provost for academic affairs, and Susan, associate provost for educational development and director of the Eberly Center for Teaching Excellence, applied their expertise, as well as the experiences of many graduate and faculty women, to create a handbook designed to give women the basics of managing a doctoral career in the engineering and science fields. Although we, in our efforts at Carnegie Mellon and beyond, are committed to working toward an institutional transformation that will create an equitable environment in which everyone can learn and work, we wrote this guide to help women navigate a still imperfect system. In addition, by highlighting issues that women in engineering and science face, we hoped to identify key areas for needed change.

Although neither Barbara nor Susan holds a doctorate in engineering or science, we have collaborated on several similar projects before and have done considerable work to promote the advancement of women in these traditionally "male" fields. In addition to our collaborative work, *Journeys of Women in Science and Engineering: No Universal Constants* (which profiles 88 contemporary women scientists and engineers), we have also authored or coauthored other pertinent works.

Barbara, an anthropologist who was formerly associate dean and director of the Center for Women's Careers at Wellesley College, has published widely on the topics of increasing women's numbers in engineering and science. Among other publications, she has contributed to *The Equity Equation: Fostering the Advancement of Women in the Sciences, Mathematics and Engineering* (Jossey-Bass, 1996) and has written "The Way Girls Learn: A Patchwork Quilt of Impressions" (which appeared in the *Indian Journal of Gender Studies,* 1996). Barbara served as an advisor to MentorNet, a service that pairs women studying engineering and science with women mentors in industry, and she is on the board of directors of Women in Engineering Program Advocates Network (WEPAN).

   Susan, a social historian, coauthored *The New Professor's Handbook: An Introductory Guide to Teaching and Research in Engineering and Science* (Anker Publishing, 1994) and has written chapters in such books as *Improving College Teaching* (Anker Publishing, 1995) and *Preparing the Professoriate of Tomorrow to Teach* (Kendall/Hunt Publishing, 1990).

   Lisa Ritter, who holds a master of arts degree in professional writing from Carnegie Mellon, joined with Barbara and Susan to bring another perspective to the book. Lisa, who has run professional development seminars for graduate students and currently edits the quarterly *Graduate Times* newsletter at Carnegie Mellon, further shaped the draft of the book and interviewed many of the women whose stories and advice are included here.

   Although this book was several years in the making, we believe it reflects an evolution that gives the book its strength because it contains the input of many women who have received (and who are working toward) Ph.D.s in engineering and science. We would like to thank these women as well as many others who have helped shape the book's content: those who gathered information for early drafts of the book; speakers from Carnegie Mellon's Graduate Student Seminar Series; friends and colleagues of the authors who have read drafts and supplied pertinent data; and others who have helped bring the book from concept to finished product.

   We hope that you will find the tips and insights that we have gathered useful as you pursue your doctorate.

Barbara B. Lazarus

Lisa M. Ritter

Susan A. Ambrose

PREFACE

A BRIEF HISTORY OF WOMEN AND THE PH.D.

Depending on your field, you may be among a very small minority of the doctoral students in your science or engineering department. But just over 100 years ago, you would not have been formally admitted at all to the doctoral programs at any university in the United States. The reasons given for not accepting women ranged from those based on bad social science to silly medical science. University administrators—and society at large—espoused the view that such education would be wasted on women since they would soon be married and busy raising children. Some "scientists" of that time even believed that studying drew away blood necessary for menstruation and pregnancy, thus hampering a woman's procreation abilities (Caplan, 1994). But even after such ridiculous arguments had been debunked, universities could still offer one final—and unarguable—excuse: there was no precedent for accepting women.

Although women at that time weren't being formally accepted into graduate school, many did find ways to get in, usually as "special" or nondegree students. For example, in 1870 Ellen Swallow Richards applied to the Massachusetts Institute of Technology for a graduate degree in chemistry, but was instead admitted as a "special student" who was seeking a second bachelor's degree (her first one was from Vassar). It was argued that the Chemistry Department didn't want its first graduate degree to be awarded to a woman (Rossiter, 1982). The difficult part, it seemed, was actually being awarded a degree for the work done. One of the most infamous examples of this discrimination is the case of mathematician Christine Ladd-Franklin, who completed her dissertation in 1882 at Johns Hopkins University but was not awarded her degree until 44 years later (Rossiter, 1982).

It wasn't until 1890 that women could officially enter graduate school in the United States, and then only a handful of universities allowed them. Over the following decades, more universities opened their doors to women doctoral students, but many barriers still remained, especially in the historically male rampart of science and engineering. This tradition of exclusion of women in these fields becomes easier to understand, however, when one examines the history of the disciplines.

Even with over a century of U.S. graduate experience under our belts, women's involvement in the academy is still relatively new, and our official acceptance into science and engineering programs infantile. In the Middle Ages, European universities were created largely to teach theology, medicine, and law—professions that were not open to women. There were a few remarkable exceptions (such as Laura Bassi, who received her doctorate in 1733 and was a physics professor at the University of Bologna), and of course, the countless—and not so famous—numbers of women throughout the ages who studied and practiced science and engineering in whatever way they could. For the most part, however, women just weren't welcome in these professions (Ambrose et al., 1997).

Science's parameters—established by Descartes and Francis Bacon—have been blamed by some scholars for alienating women. Descartes pronounced that objectivity in science required a separation of logic from emotion (the "mind–body problem"). Bacon "declared that the purpose of science was for man to conquer and control nature for his [own] benefit" (Ambrose et al., 1997). Some writers hold that these tenets—and the scientific culture built upon them—have served to exclude women from the official scientific community in the past.

Engineering was even more exclusionary, due to its inception from a military model. In the 1400s, the word "engineering" (from the Latin word for "to contrive") was used to describe the design of devices for warfare. Thus, an engineer was a man who employed skills to build machines of war. Even when engineering took on the qualifier "civil" to denote its nonmilitary applications, it remained a man's domain.

Although formal science and engineering education has long been the realm of men, even nonmilitary male students weren't able to formally study engineering in the United States until 1824 when the Rensselaer School (now Rensselaer Polytechnic Institute) was established (Grayson, 1977). The first doctorate ever earned in the United States wasn't awarded until 1861, when Yale granted one to a man in science.

This brings us back to the late nineteenth century, when women started to sneak in the back door of graduate programs and slowly but surely began to break down the barriers that had kept them out.

This is just a brief look at how women fit into the bigger picture of the doctorate in science and engineering. This background, however, gives us a basis on which to examine the current situation for women seeking their doctorates in those fields.

WHAT IT'S LIKE NOW

The National Science Foundation reports that women are currently earning the majority of bachelor's and master's degrees overall, but in 1998 earned just 40% of the doctorates in science and only 13% of those in engineering (National Science Foundation, 2000). More than a century after Christine Ladd-Franklin was denied a doctorate, these and other recent numbers make it clear: many women scientists and engineers face challenges throughout their careers simply because they are women. These challenges are compounded for women of color, lesbians, differently abled, and economically disadvantaged women. Most fields are male dominated at every level, and female mentors and role models are few. Doctoral work is certainly no exception, and it is only a first hurdle along the journey to a fulfilling career.

For the last 20 years, the United States has faced a critical shortage of scientists and engineers with doctoral degrees in a number of fields. Even in areas where absolute numbers are not an issue, there are real equity concerns. The number of women working and studying in the scientific and technical fields is markedly lower than the number of men. To help all individuals reach their potential and to remain competitive into the next century, it is fundamental that the United States recruit from our entire population and that we support greater numbers of women in scientific and technical Ph.D. programs.

Table 1 Number of Science and Engineering Doctorate Recipients, by Gender and Subfield, 1998

Subfield of Doctorate	Number of Doctorates		
	Total*	Men	Women
PHYSICAL SCIENCES	6,739	5,104	1,600
MATHEMATICS	1,177	872	297
Applied Mathematics	265	203	61
Algebra	75	57	18
Analysis & Functional Analysis	130	105	25
Geometry	54	39	15
Logic	16	11	5
Number Theory	46	39	7
Mathematical Statistics	204	141	62
Topology	65	52	13
Computing Theory & Practice	18	15	3
Operations Research	17	11	6
Mathematics, General	163	118	39
Mathematics, Other	124	81	43
COMPUTER SCIENCE	923	763	157
Computer Science	817	696	118
Information Sciences & Systems	106	67	39

(Continued)

Table 1   *(Continued)*

	Number of Doctorates		
Subfield of Doctorate	Total*	Men	Women
PHYSICS & ASTRONOMY	1,584	1,354	223
Astronomy	91	69	22
Astrophysics	117	93	24
Acoustics	18	12	6
Chemical & Atomic/Molecular	99	86	13
Elementary Particles	173	162	11
Fluids	26	26	0
Nuclear	92	81	11
Optics	104	86	18
Plasma & High-Temperature	55	53	2
Polymer	24	20	4
Solid State & Low-Temperature	313	276	37
Physics, General	190	158	29
Physics, Other	282	232	46
CHEMISTRY	2,217	1,510	695
Analytical	384	238	146
Inorganic	287	203	84
Nuclear	5	4	1
Organic	597	437	160
Medicinal/Pharmaceutical	115	68	46
Physical	278	201	77
Polymer	123	83	40
Theoretical	41	31	10
Chemistry, General	286	187	88
Chemistry, Other	101	58	43
EARTH, ATMOS., & MARINE SCI.	838	605	228
Atmospheric Physics & Chemistry	38	31	7
Atmospheric Dynamics	24	16	8
Meteorology	25	19	5
Atmos. Sci./Meteorology, General	22	13	8
Atmos. Sci./Meteorology, Other	16	14	2

Subfield of Doctorate	Number of Doctorates		
	Total*	Men	Women
Geology	171	131	40
Geochemistry	58	35	22
Geophysics & Seismology	106	85	21
Paleontology	23	16	7
Mineralogy, Petrology	14	9	5
Stratigraphy, Sedimentation	24	20	4
Geomorphology & Glacial Geology	20	12	8
Geological & Related Sci., General	13	8	5
Geological & Related Sci., Other	40	33	7
Environmental Science	73	43	30
Hydrology & Water Resources	35	27	8
Oceanography	94	64	29
Marine Sciences	18	15	3
Misc. Physical Sciences, Other	24	14	9
ENGINEERING	5,919	5,108	769
Aerospace, Aeronautic., Astronautic.	242	227	14
Agricultural	73	68	5
Bioengineering & Biomedical	207	157	50
Ceramic Sciences	24	22	2
Chemical	667	542	119
Civil	587	498	83
Communications	40	34	6
Computer	210	187	22
Electrical, Electronics	1,343	1,206	127
Engineering Mechanics	86	73	13
Engineering Physics	15	12	3
Engineering Science	50	42	8
Environmental Health Engineering	63	46	17
Industrial/Manufacturing	227	187	39
Materials Science	482	404	75
Mechanical	936	849	79
Metallurgical	59	51	7

(Continued)

Table 1   *(Continued)*

Subfield of Doctorate	Number of Doctorates		
	Total*	Men	Women
Mining & Mineral	21	17	4
Nuclear	97	86	10
Ocean	29	29	0
Operations Research	62	47	15
Petroleum	48	42	6
Polymer/Plastics	59	44	15
Systems	68	59	9
Engineering, General	30	23	3
Engineering, Other	194	156	38
LIFE SCIENCES	8,540	4,640	3,876
BIOLOGICAL SCIENCES	5,848	3,298	2,533
Biochemistry	798	448	349
Biomedical Sciences	184	101	79
Biophysics	166	119	47
Biotechnology Research	12	8	4
Bacteriology	13	9	4
Plant Genetics	40	22	18
Plant Pathology	18	10	8
Plant Physiology	61	33	28
Botany, Other	113	58	55
Anatomy	35	27	8
Biometrics and Biostatistics	75	35	39
Cell Biology	299	145	154
Ecology	292	177	114
Developmental Biology/Embryology	127	66	61
Endocrinology	30	16	14
Entomology	138	101	37
Biological Immunology	245	130	115
Molecular Biology	741	414	324
Microbiology	384	214	169
Neuroscience	412	244	168
Nutritional Sciences	137	42	95

Subfield of Doctorate	Number of Doctorates		
	Total*	Men	Women
Parasitology	15	9	6
Toxicology	156	95	61
Human & Animal Genetics	196	105	91
Human & Animal Pathology	91	56	35
Human & Animal Pharmacology	256	133	121
Human & Animal Physiology	258	158	100
Zoology, Other	111	68	43
Biological Sciences, General	217	133	82
Biological Sciences, Other	228	122	104
HEALTH SCIENCES	1,500	488	1,006
Speech-Lang. Pathology & Audiology	95	20	74
Environmental Health	54	37	17
Health Systems/Services Admin.	63	27	36
Public Health	157	49	107
Epidemiology	166	54	112
Exercise Physiology/Sci., Kinesiology	129	80	49
Nursing	399	17	380
Pharmacy	156	79	75
Rehabilitation/Therapeutic Services	33	11	22
Veterinary Medicine	48	30	18
Health Sciences, General	17	5	12
Health Sciences, Other	183	79	104
AGRICULTURAL SCIENCES	1,192	854	337
Agricultural Economics	155	115	40
Agricultural Business & Management	2	2	0
Animal Breeding & Genetics	18	12	6
Animal Nutrition	45	32	13
Dairy Science	10	8	2
Poultry Science	11	8	3
Fisheries Science & Management	30	22	8
Animal Sciences, Other	60	44	16

(Continued)

Table 1   *(Continued)*

Subfield of Doctorate	Number of Doctorates		
	Total*	Men	Women
Agronomy & Crop Science	96	77	18
Plant Breeding & Genetics	69	55	14
Plant Pathology	66	42	24
Plant Sciences, Other	37	23	14
Food Engineering	13	10	3
Food Sciences, Other	153	87	66
Soil Chemistry/Microbiology	27	20	7
Soil Sciences, Other	74	53	21
Horticulture Science	60	47	13
Forest Biology	20	14	6
Forest Engineering	2	2	0
Forest Management	27	18	9
Wood Sci. & Pulp/Paper Tech.	25	21	4
Conservation/Renewable Nat. Res.	25	14	11
Forestry & Related Sci., Other	69	53	16
Wildlife/Range Management	55	41	14
Agricultural Sciences, General	8	7	1
Agricultural Sciences, Other	35	27	8

*Grand totals include 174 doctorate recipients whose gender was unknown and 20 doctorate recipients whose doctoral field was unknown.

NOTE: Field groupings may differ from those in reports published by federal sponsors of the Survey of Earned Doctorates.

Excerpted from Doctorate Recipients from United States Universities: Summary Report 1998 (see Sanderson et al. in Bibliography).

See http://www.nsf.gov/sbe/srs/srs00410/tables/taba1.xls for data on other disciplines.

Source: NSF/NIH/NEH/USED/USDA, Survey of Earned Doctorates

Why aren't more women earning their Ph.D.s in science or engineering? Why aren't women with degrees remaining in academic careers? The answers are complex but have a great deal to do with the nature of doctoral education, the atmosphere, and largely unconscious attitudes and cultural biases in many graduate programs in the sciences and engineering. In our culture, we presume that math ability is male-gendered and innate. In other cultures, everyone is assumed to be able to learn math, so it is taught to everyone. This disparity in presumed ability to do math can become a source of low self-esteem in women raised in the American culture. Even if they are "good in math," they may see all men (and women from other cultures) as better in the subject than they because they lack the self-confidence that comes from being nurtured in math education. In countries where high school courses in math and science are required, women do better in science (Schiebinger, 1999).

Doctoral work, particularly in science and engineering, uniquely challenges women and, in many cases, promotes the success of men. A number of people in higher education still cling to the outdated model that only white men have the ability to be successful scientists and engineers, that women are best suited to careers in the humanities and social sciences, and that women just aren't tough enough to succeed in technical fields.

For women to survive in graduate school—and to emerge ready for careers in the academy, industry, or the public sector—they must overcome stereotypes and hidden barriers. Women need to learn how to maneuver in a predominantly male graduate school environment, how to think like academics, and how to be politically astute.

This book is designed to unravel some of the mystery around graduate school programs in science and engineering and to provide strategies for succeeding. Each chapter covers a different aspect of graduate school, from identifying funding sources to writing the dissertation to looking for a job. The guide focuses on the emotional and social challenges women may face, and it offers practical suggestions and advice for surviving and thriving in graduate school, not because all

---

WHAT IT TAKES

"When I think about what it takes for women to be successful as scientists, four categories come to mind into which most successful women scientists fall: some are oblivious to negative feedback; some are stubborn and the more they hear 'you can't do this,' the harder they try; others are very creative and take unusual paths in both their careers and personal lives; and a few are simply incredibly brilliant. Most women fall into the first three categories, and there is room for all types of personalities. I consider myself part of the 'stubborn' category. I've always found science fun and stimulating. For me it is important to remember the wonderful aspects of it even when I experience unpleasantness."

Debbie C. Crans
Associate Professor of Organic Chemistry
Colorado State University
(Ph.D., Organic Chemistry, Harvard University)

From *Journeys of Women in Science and Engineering: No Universal Constants*

women will face any or every problem, but because those who do are most in need of advice. The guide covers the practical side of graduate school including course work, choosing an advisor, and funding concerns. It also discusses common problems and concerns many female graduate students report. The bibliography contains our sources and other helpful publications.

Interspersed throughout are personal stories and advice from current female graduate students based on their experiences in graduate school and from women engineers and scientists from across the United States. All of the stories included in the guide are from real people in actual situations, but in most cases the names and departments have been changed to keep their identity confidential.

We hope you will use our guide as a resource when considering graduate school or as a handbook to help you thrive. And never forget: the path is worth it. Women who work in these careers enjoy relatively high standing and salaries, job security, and mobility. But most importantly, they work at what they love and have a chance to make a difference.

Table 2   Timeline

Early 1800s	Technical institutes and land-grant colleges educate engineers and other technicians who could feed America's burgeoning economy. Higher education was becoming less elite and more practical (Boyer, 1990; Glassick et al., 1997).
1836	Georgia Female College at Macon opens and is the first institution in the United States to grant degrees to women. Required courses included astronomy, chemistry, and mineralogy (Ambrose et al., 1997).
1837	Oberlin College is the first college to admit women into its regular program (Selby, 1999).
1861	The first U.S. doctorate is awarded at Yale (in science) (Davis et al., 1996).
1870	Ellen Swallow Richards applies to MIT as a graduate student in chemistry, but is accepted instead as a "special" student pursuing a second bachelor's degree (Rossiter, 1982).
1876	Johns Hopkins University is founded, mainly to offer doctoral degrees (Boyer, 1990).
1882	Mathematician Christine Ladd (later Ladd-Franklin) submits her dissertation at Johns Hopkins University, but is refused a degree by the Board of Trustees (Rossiter, 1982).
1890– 1892	Women are accepted as graduate students by Yale, the University of Pennsylvania, Columbia, Brown, Stanford, and the University of Chicago. Because of peer pressure, other universities soon followed suit and, by the early 1900s, women were accepted into most graduate schools (Rossiter, 1982).
World War I	A boom in employment occurs for women in engineering and science. Although these women had proven that they could do "men's work," returning soldiers took back their jobs and the women engineers and scientists were sent "home."
early 1920s	Harvard Medical School appoints Alice Hamilton as its first woman faculty member. Hamilton, the best qualified candidate Harvard could find in industrial medicine, was not allowed to use the faculty club or march in the commencement procession (Glazer and Slater, 1987).

Figure 2 In the 1920s, women earned 15% of all doctorates (Lomperis, 1990) and 12% of science doctorates (Vetter, 1984). (Photo credit: Courtesy of Carnegie Mellon University Archives.)

1926   Christine Ladd-Franklin is awarded her Ph.D. in mathematics from Johns Hopkins University—44 years after she earned it (Green and LaDuke, 1987).

1940   Thirteen percent of all doctorates are awarded to women (Chamberlain, 1988).

1944   G.I. Bill is first authorized. Almost 8 million veterans use the bill to attend college, but only 3% of the veterans enrolled in college in the 1940s were women (Hollenshead, 1998).

1947   President Truman creates the President's Commission on Higher Education, which seeks to shift emphasis from educating the elite to educating the masses (Boyer, 1990).

1950   The National Science Foundation is created to "promote the progress of science."

1950s   Fewer women receive Ph.D.s than in the 1920s (Hollenshead, 1998).

1960   Women earn 2.6% of engineering, physics, and mathematics doctorates and 8.8% of life sciences doctorates (Lomperis, 1990).

Figure 3 In the 1950s, fewer women received Ph.D.s than in the 1920s (Hollenshead, 1998). (Photo credit: Courtesy of Carnegie Mellon University Archives.)

(Continued)

Table 2   (*Continued*)

1960s	The Civil Rights Act and the Federal Women's Program outlaw discrimination in employment.
1970	Women earn 3.6% of engineering, physics, and mathematics doctorates and 13% of life sciences doctorates (Lomperis, 1990).
1980	Women earn 9% of engineering, physics, and mathematics doctorates and 25.9% of life sciences doctorates (Lomperis, 1990).
1989	Women earn 28.1% of all science and engineering doctorates (National Science Foundation, 2000).
1989	Harvard tenures its first woman chemistry professor (Cynthia Friend) and in 1992 tenures its first woman physics professor (Melissa Franklin).
1998	Women earn 34.3% of science and engineering doctorates (National Science Foundation, 2000).
1998	Microbiologist Rita Colwell is appointed the first female head of the National Science Foundation.

ACKNOWLEDGMENTS

The authors are very grateful to the following for their wisdom, their time, and their graciousness in providing information, experience, and feedback for this book: Linda Babcock, Jack Beuth, George Bekey, Anita Borg, Susan Burkett, Lin Chase, Cinda-Sue Davis, Julia Deems, Denice Denton, IEEE reviewers and staff, Nancy Klancher, Sara Majetich, Deirdre Meldrum, Linda Melville, Indira Nair, Illah Nourbakhsh, Lynn Philibin, Tresa Pollock, Sarah Rajala, Teodora Rutar Shuman, William Slye, Laurie Weingart, and, in particular, the many graduate women who have shared their stories with us.

We offer special thanks to the following past and current Carnegie Mellon graduate students for their research assistance on earlier drafts of the book: Carol Goldburg (Ph.D., Graduate School of Industrial Administration), Jade Goldstein (Ph.D. candidate, Language Technologies Institute), John Hinshaw (Ph.D., History), Mark Kantrowitz (Ph.D. candidate, Computer Science), Kathleen Kindle (M.A., Professional Writing), Margaret Kanipes (Ph.D., Biology), Margaret Mc-Caffrey (Ph.D., English/Rhetoric), Francesmary Modugno (Ph.D., Computer Science), Brett Molotsky (Ph.D., English/Literary and Cultural Theory), Seth Ruffins (Ph.D., Biology), and Lara Wolfson (Ph.D., Statistics).

This book is the result of many years of research, teaching, and experience. As is the case in many long-term enterprises, it has benefited from collaborations with numerous students and colleagues. We have done our best to properly cite and verify their contributions. If we have made any errors in this process, we welcome having them brought to our attention so that we may make accurate attributions in future printings of this work.

Barbara B. Lazarus

Lisa M. Ritter

Susan A. Ambrose

CHAPTER 1

INTRODUCTION

WHAT IS DOCTORAL EDUCATION?

- The Ph.D. Program
- The Demands of Doctoral Work
- Why Some Women Find Doctoral Work *More* Than Demanding
- How to Make It

The fundamental purpose of a doctoral program is to help you become a professional who generates ideas within an extremely specialized field. As an undergraduate, you probably spent most of your time reading and writing about established theories and research. You were required to learn existing knowledge. By contrast, doctoral study will challenge your ability to learn in a completely different way. It focuses study in one discipline, and it requires that you conduct original research and formulate your own theories. Whether explicitly stated or not, as a doctoral student you will be expected to create new knowledge.

Graduate school should also be considered an intensive training process. During the course of your doctoral work, you become an expert in your field. Many of you will learn how to become faculty members; others, professional researchers and practitioners in the private sector, the public sector, or other nonprofit organizations. The work you do now will begin to establish your credibility as an authority on your chosen subject.

The Ph.D. Program

The highest degree awarded by academic graduate programs is a doctorate. Although there are many legitimate "doctoral" degrees—such as the J.D., Juris Doctor for lawyers; Ed.D., Doctor of Education; Pharm.D., Doctor of Pharmacy; and Th.D., Doctor of Theology for religious ministers—the recognized research degree

1

Indira Nair

AN INTELLECTUAL HIGH

"The Ph.D. is a challenging and tough endeavor for everyone because of the exploration into uncharted places of knowledge. This is also what makes it inviting, exciting and worthwhile, because one is able to discover one's own intellectual strengths as well as the way one's curiosity develops during this time. There is no other instance in which one can feel quite the same 'intellectual high' and sense of accomplishment that one's own Ph.D. work gives, so it is important to understand and learn to navigate the process."

Indira Nair
Vice Provost for Education and Associate Professor of Engineering
and Public Policy, Carnegie Mellon University
(Ph.D., Physics, Northwestern University)

for the liberal arts, the sciences, and engineering is the Ph.D. (or, in some institutions, the Sc.D.).

A Ph.D. has the following basic requirements:

- Residency in an approved program of study.

- Proficiency in a prescribed body of knowledge within a specific field.

- Proven ability to perform independent investigation on a significant problem within a specific field, the evidence of which is the presentation and defense of a dissertation on an original piece of research.

Some doctoral programs incorporate a master's degree into the course of study for a Ph.D. (Some require completing a master's degree before entering the program.) A student enrolls in the program after undergraduate school, receives a master's degree in the first year or two of study, and then continues working toward the Ph.D. In many departments, however, the Ph.D. process is separate from the M.S. process, and the M.S. is not necessarily a prerequisite for the Ph.D.

The typical degree program lasts approximately four to six years and incorporates courses in your chosen field of study with research. You build on your course work and original research to write a dissertation that is presented and defended before a faculty committee. This is a much abbreviated description of what happens; greater detail is presented throughout this guide.

The Demands of Doctoral Work

You've probably already heard how tough doctoral work is—how demanding the schedule is, how long it takes, and how the workload gets heavier with each passing term. All of this is true—and then some. In fact, doctoral work may be one of the most challenging experiences of your life to date. It can also be one of the most rewarding. If you know what to expect and if you have a solid knowledge base and skills for survival, you will be better prepared to meet its challenges and reap its rewards.

You have to understand that there are few standard rules in graduate school. Requirements and expectations vary among schools, departments, and disciplines. Some institutions have no central graduate school that handles admission and administrative policies and procedures; each college, department, or program within the university may have its own requirements for admission and completion of a graduate degree. Other schools, often larger universities, have formal graduate schools with requirements for admission and graduation that apply to every graduate student. Individual departments and programs may have their own requirements in addition to the graduate school's; however, for admission and graduation, each student must meet the minimum standards.

Creating new knowledge is the most exciting thing about a Ph.D.; your Ph.D. program is designed to help you develop the skills necessary to create that knowledge. The first few terms of your program, through classes and meetings with your advisor, help you discover where your research interests lie. Your course work and qualifying exams are designed to provide you with the theory and research tools necessary to develop "ownership" of knowledge. Finally, through planning and writing a dissertation, you can begin to outline and explore research questions of your own.

Apart from academic requirements, doctoral work can also place enormous demands on your self-confidence. Graduate courses are typically much harder than undergraduate courses. Overwhelming duties like developing a research study, teaching classes, and writing and defending a dissertation can all place great stress and strain on your mental faculties. Doubting yourself and your ability to survive is perfectly natural. You may ask yourself questions like, "Can I discipline myself to write a major paper? Will I ever create something new? Do I have what it takes to make a significant contribution in my area? What do my advisors and committees think of me?" Recognizing that stress and self-doubt are a natural part of any significant experience, including graduate school, tends to help.

OBLIVIOUS AND CONFIDENT

"My graduate class was small, but about one-third of it was women. The department was very supportive, and it was a fabulous place to be a graduate student. There were some people who didn't think women belonged, but the other women graduate students and I simply avoided them. I must say that I was pretty oblivious to the attitudes of others about my being a woman in science. I guess that was a blessing because I never felt like I didn't belong or shouldn't be pursuing something that I loved. I learned early on that it's a very good ploy to act confident even when you're not because then people perceive you as confident, and that makes a big difference."

Lydia Villa-Komaroff
Associate Vice President for Research Administration
and Professor of Neorology, Northwestern University
(Ph.D., Cell Biology, Massachusetts Institute of Technology)

From *Journeys of Women in Science and Engineering: No Universal Constants*

Why Some Women Find Doctoral Work *More* Than Demanding

There is no doubt that graduate school is extremely demanding for everyone. However, it does pose certain challenges that affect women more than men. These challenges can be more complicated for women of color, lesbians, differently abled, or economically disadvantaged women.

Graduate school is a relatively recent option for women, especially those interested in the sciences and engineering. Traditionally, white men pursued doctorates, while most women were held by societal and cultural precedent to a bachelor's degree at best. Today, many women graduate students meet "invisible barriers" within the graduate school system. Many of these barriers are also invisible to men in the environment, and they may unwittingly participate in perpetuating the system. When questioned about whether they're supportive of women in their departments, these men may answer "yes," but unknowingly they may have access to information and contacts their female counterparts do not. The barriers—and the system—are unknown to most women because of their status in the environment. In fact, many women in engineering and science departments around the country report that graduate schools still operate like an "old boys club." Whether by design or default, women may find that they are not privy to the secrets of successfully maneuvering the system, and too often they blame themselves for faults that are part of that invisible system. In some cases, women are welcomed with open arms to academic engineering departments but are *still* isolated by a system that was put in place well before their arrival. Fortunately, current research on the graduate school experience for women has identified many of these invisible barriers and has presented ways to cope with them—as we hope to do in this guide.

One well-documented barrier is finances. Historically, some women in the sciences and engineering did not receive comparable financial support to men, including grants and postdoctoral positions. In the short term, women may face greater financial and professional strain. Some women have been forced to drop out of Ph.D. programs due to a lack of funding. For many women, financial burdens coupled with an absence of female colleagues and mentors can make the entire experience too difficult and too costly at a personal level. In the long term, inequities can erode the personal strength and self-confidence of female graduate students and can cause them to lose faith in the system. More recently, many colleges and universities have recognized the limited support for women in graduate school and have begun to revise their funding policies.

Women still encounter difficulties because of stereotypes that depict them as intellectually inferior, stressed-out, or wanna-be mommies, but not workers. As a result, faculty and fellow students may unconsciously gauge their views of fellow female classmates in line with limited expectations. Too often women encounter subtle forms of resistance that challenge their personal self-confidence and limit their access to university community resources such as mentoring, collaboration, and informal exchanges vital to academic success.

Although it may seem that women in doctoral programs have already made the choice of career with or without family, the lines are usually not so clear. Some are older and already have families when they begin school—and the majority of the responsibility for housework and child care remains with them. The majority of women still have to cope with greater demands on their personal time than their male colleagues. Female graduate students who also happen to be wives and mothers may encounter subtle forms of discrimination for not being "serious students" and for not being able to completely commit themselves to their graduate work 24 hours a day. They also experience more stress due to expectations at home.

How to Make It

These and other challenges make it that much harder for many women to succeed in graduate programs. As a woman in science or engineering, you've probably already experienced some of these problems during your undergraduate years, although many women don't recognize then until they reach graduate school. You know how the system operates, and, obviously, your enthusiasm for your field has carried you this far. The belief in yourself and your ability to succeed, as well as your passion for learning and study, will help you make it through graduate school. When you enjoy what you do, working hard to succeed is worth it.

Women in graduate school in science and engineering today are less likely to be the pioneers they were a generation ago. Although you still face problems simi-

A RATIONAL OUTLOOK

"I wanted to do engineering because it suited me—it was very rational and I liked its exactness. I have always enjoyed problem solving, and was always intrigued and excited about reaching some sort of conclusion.

"The idea of graduate school came to me when I realized that I was a better student as an undergrad than most of the other students, and that I could probably contribute to society more by getting a Ph.D. I also realized that I had the potential to do much more than calculate the flow rate in multiple pipe systems!

"When I was an undergraduate student in Belgrade, Yugoslavia, about 30–35% of the students in the mechanical engineering program were women. I do not remember any issues of gender differences—we all felt like pretty good friends. When I entered graduate school in the United States, I continued to take a proactive approach focused on my studies. I really did not feel that being a woman was going to be to my disadvantage. That positive attitude was and is crucial to bringing my studies to completion.

"I think the fear of discrimination that is imbedded into women is exaggerated. Although some women do experience true discrimination, some women believe that being a woman is advantageous. My overall experience is somewhere halfway between to two extremes. An engineering doctoral program is demanding in its essence, and I believe that thinking that hard times are inevitable could lead to bad judgment. It is easy to confuse gender discrimination with problems such as professors or fellow students lacking people skills.

"A word of advice: be proactive, stay oriented toward achieving your career goal, and every day take a step forward."

Teodora
doctoral student, mechanical engineering

lar to those who came before you, and the situation in most graduate schools is far from perfect, it is improving. The encouraging news is that, as a woman in graduate school today, you will be able to experience and influence changes that your predecessors never imagined.

In this guide we share successful strategies developed by other students for surviving and thriving in their doctoral work. We also have included many suggestions for developing the knowledge and skills to prepare you for your life's work. Graduate school will truly test your commitment, energy, and patience. Probably no other situation in your life to date will be as overwhelming—or rewarding.

CHAPTER 2

THE JOURNEY

"Our best hope for the future are women who don't see the ceiling but the sky."

Claudia Goldin
First tenured woman in Economics at Harvard
From "Why Women Don't Hit the Top," *Fortune,* July 30, 1990

THE ADVISOR

- What Is an Advisor?
- Choosing an Advisor
- Making the Advising Process Work
- Other Sources of Support and Advice
- What If It's Not Working?
- How Can You Make the Change?
- Problems Women May Encounter in the Advisor/Advisee Relationship
- Getting the Support You Need
- The Importance of Having a Mentor

What Is an Advisor?

Your advisor is an indispensable resource. She or he suggests what courses you should take, directs your program of study, supervises your research, and chairs your dissertation and oral exam committees. At best an advisor is your confidant, mentor, sponsor, and major advocate in the department now and later in your career. Your advisor can be the launching pad for your career by giving you the opportunity to work on well-funded projects, by introducing you to contacts and influential people in your field, and by providing opportunities to exercise your skills outside of the university. Ideally, the relationship between you and your advisor will progress during your time in graduate school from that of teacher and research director to that of mentor and professional colleague.

TIP

Remember that the most important relationship in graduate school is between you and your advisor. Your success as a student depends on it; and the course of your career may depend upon it as well.

Choosing an Advisor

Advising relationships develop in a variety of ways. In some cases, new graduate students are assigned to advisors based on their field of study and research interests. As we will see later, advisors may assume financial responsibility for their advisees as early as the first year.

Other programs allow graduate students to choose their own advisor from the beginning. These students are usually expected to choose an official advisor at some point early in the first few years of their graduate careers. When making this crucial decision, carefully consider the following:

- If possible, choose a faculty member who shares your interests, philosophies, and general world view. You want to choose someone with whom you feel comfortable and who will give you the individual support you may require. Try to get acquainted with several faculty members in your department before making a definitive choice. Ask the opinions of other graduate students.

- Make sure you can get along with a potential advisor on both a professional and personal level. You will be spending a great deal of time together—in most scientific fields, five or six years—so you need to work together comfortably. You and your advisor should be able to communicate openly and honestly. This doesn't mean that you and your advisor should be best friends, just that you can work together professionally.

- Ask around the department about potential choices. Find out how a certain faculty member has interacted with other graduate students, if she or he has previously served as an advisor, and what her or his philosophy is on the "graduate school experience." It may be to your advantage to determine whether your potential advisor is well-connected in terms of financial support and outside collaborations.

- See if you can find out whether your potential advisor will be on campus long-term—presumably for the next four to five years. If there is a possibility that he or she will move, think about your options: would you move along

---

CHOOSING THE RIGHT ADVISOR

"When I was looking for an advisor, I was looking for someone who could help me practically and emotionally. I need a little more feedback than most people, especially positive feedback. There are times when I need to say, 'I need you to check up on me once a week.' And there are times when I'll say, 'I need to be left alone for a month.'

"I talked to other graduate students before I started looking at research areas. There were 15 faculty members—9 or 10 that I considered as advisors. Some I crossed off the list right away because I wasn't interested in the area. And there were others who couldn't provide the kind of emotional support I needed. I worked with people on small projects that were interesting and educational and gave me an opportunity to get to know potential advisors. That was how I met my advisor.

"Everything is really working out well. My advisor is willing to look at my talents and at me as an individual."

Christine
doctoral student, civil engineering

---

with your advisor, or would someone else in the department be able to take you on as an advisee?

- Think about the issue of junior or senior faculty status. Most senior faculty members are presumably better connected both within and outside the university. On the other hand, some junior faculty members may be able to spend more time with you, but in some instances may be too busy launching her or his own career to advise you properly. Many graduate students advise optimizing the advisor relationship while satisfying the research area.

- If your first choice for an advisor is not available or is not suitable for one reason or another, or if you're torn between two choices, you may still be able to have that person on your dissertation committee or work with him or her in some other capacity.

Making the Advising Process Work

Remember, you are half of the advisor/advisee relationship. To get the most out of it, you need to do your part. It is your responsibility to seek out a potential advisor who will help and support you according to your needs. You should be aware of what is required of you as well as of your advisor. As a graduate student, you have the responsibility to your advisor to:

- Express your seriousness and commitment to completing your degree by working hard.

- Meet regularly and talk with your advisor, preferably on a mutually agreed-upon schedule.

- Build your advisor's trust by seriously listening and considering her or his advice and constructive criticism. Be honest and discuss all issues, including those that bother you.

- Share your future plans.

- Be self-motivated.

- Continually expand your knowledge and test new ideas. Attend seminars and colloquia held on campus. Learn to read the literature in your field.

Your advisor is, above all else, the person responsible for molding you into a Ph.D. There is often the temptation to develop a close friendship in which you share your personal problems; after all, you spend a great deal of time together. Many students advise against getting "too close" because your advisor should view you as a committed, independent, and capable young professional—not as a child, friend, or patient. You can be on friendly and honest terms with each other, but as a doctoral student, you also need to demonstrate and prove your independence.

Other Sources of Support and Advice

Although officially you have one advisor—the person who heads your thesis committee and supervises your academic program—you don't have to limit yourself to one source of advice and support. Many graduate students develop a series of advising and mentoring relationships with faculty on and off their committees. Some women find it useful to seek out many friendly supporters, whether or not they are involved with their department. It's unlikely that you will receive all the support you require from just one person, so having more people to count on can be better.

Don't overlook other graduate students, both women and men, as sources of support and advice. Although faculty members are important to your professional development, graduate students, within and outside your department, can be invaluable. Your peers now will be your professional colleagues in the future; in some cases they may be your best source of emotional support. As several graduate students advise: Guard against being labeled too emotional by faculty; find graduate students to share your feelings with.

What If It's Not Working?

If you really aren't receiving the support you need from your advisor, you may want to consider choosing someone new. It's been done before. Some graduate students change advisors and topics; others even switch to completely different programs!

Depending on issues like financial support, your field of study, and your department, changing your advisor can be simple or it can be extremely complicated. The choice really depends on you. If you're thinking about making a change, consider the following advice from graduate students who have already done it:

- Find out about funding. If your primary financial support is tied to your advisor, making a change may not be possible.

- Don't decide to change advisors in the heat of an argument. Sit down with someone you trust, identify the problems you have, and then make a rational decision.

- Be honest with yourself; there is no perfect advisor who can give you everything you're looking for. Ask yourself if you can accept her or his limitations. If not, make a change.

- The essential question is: Can this person help me make it through graduate school and into a career? Unless you have a good alternative in mind, it might not be worth your time looking for the perfect fit.

- If you do decide to change, make the transition as smoothly and professionally as possible.

How Can You Make the Change?

If you realize early on that you and your advisor aren't compatible, you may be able to back out of the relationship gracefully. Chalk up your decision to a lack of focus.

In some ways, the most convenient "excuse" for switching advisors is a change in your research interests, although, as previously mentioned, you should check into your funding first. If your present advisor is financially responsible for you, you need a new advisor who is able to continue your funding.

Changing advisors becomes increasingly difficult when you are personally uncomfortable with an advisor who is inarguably the most appropriate person given your academic interests. You have to decide which is more important to you: working with someone you like or your academic and professional development. Of course, if your advisor is deliberately discriminating against you or acting out personal biases, you may simply have to make a change.

The farther along you are in your graduate program, the more difficult it is to change advisors. You have had a lengthy relationship with this person, and, if you leave, you can expect some feelings of resentment.

Carefully consider why you think a change is necessary. If you are changing your research focus, changing advisors is understandable. If you want to change because the personal relationship between you and your present advisor is problematic, the situation is very different. It is worth talking through your real options with your informal network of advisors, mentors, and friends before acting.

Some graduate students have found it easier to keep their official advisor and rely on other committee members for extra support. However, don't assume that these individuals know your needs or are prepared to fill the void left by your present advisor.

No matter what stage you are in in graduate school, try to handle changing advisors quietly. You still need your former major advisor in your corner. She or he may feel rejected or may agree with your decision. In either case, try to maintain a good relationship because your previous advisor will talk with your new advisor about you and your academic standing.

When you find a potential advisor who is also interested in working with you, keep the discussion of the problems with your previous advisor at a professional

---

RETHINKING A CHANGE

"Based on what I'd heard from many friends and acquaintances, I had always thought that an advisor is a mentor, guardian angel, and close friend all rolled into one. Therefore, it was a bit of a shock to me when, during a casual conversation sometime during the third year of my Ph.D., my advisor remarked that he hadn't got along well with his thesis advisor and didn't like him very much. I couldn't imagine how anybody could continue working with an advisor with whom they didn't see eye-to-eye.

"It was a couple of months after this conversation that my advisor and I had a major disagreement. It led to a lot of bitterness for me, because I realized that my advisor and I did not share the same ideals. I eventually thought of leaving the group that I was working in.

"Then I went to talk to another faculty member about research possibilities with her. We had a long conversation during which she told me almost nothing about her research. But she gave me one of the most important pieces of advice that I have ever received. 'If you decide to switch advisors and run away from your difficulties now,' she said, 'later on in life it'll always be much easier to run away from anything. First you'll leave your research program, next you'll decide to leave graduate school!' I rethought my decision to change my advisor after that.

"I continued working in the same group, but the resentment still stayed inside me. It took me about a year to realize that my attitude was making it difficult for me to work and trust my advisor. A bit of thinking made me see that except for the one big disagreement, my advisor had always been a very nice person. We had lots of communications problems, but one fine day, I just decided to let go of the resentment completely and forget about the whole thing. Now, six months later, I still can't believe how pleasant graduate school is and how much more inspired I am to work and learn. It amazes me that I now like and admire my advisor so much and that we get along so well."

Diana
doctoral student, chemistry

level. Don't risk your reputation; simply establish that your previous relationship did not meet your needs.

Problems Women May Encounter in the Advisor/Advisee Relationship

Women in science and engineering sometimes find it more difficult to find a supportive advisor than their male counterparts. More than likely, there are very few women in your department, and at least some of the potential male advisors in your department may not understand your needs simply because their experiences have been different.

For some women, one of the criteria for choosing an advisor is whether or not that individual can provide the emotional support and intellectual reinforcement they may need. On the other hand, the student cannot or will not share her concerns because she fears confirming the suspicions that women aren't good enough or tough enough to make it in graduate school. Also, many women avoid the very contact that might help them, such as asking questions or requesting help with difficult material, because they fear being perceived as weak or stupid. Often advisors either overtly or covertly minimize these feelings, and many women become frustrated and disappointed with the person they should trust the most.

In a science and engineering environment that is still influenced and controlled by men, some women have a difficult time sustaining the belief that they, too, have a place in graduate school. The erroneous assumption persists that women make it through graduate school because they receive extra attention and because their standards for completion are lower. Indeed, some women feel like "impostors" who don't belong in graduate school. Unfortunately, these biases also shape the expectations and responses of some faculty members. Even when faculty members do not share these biases, the professional and historical nature of the advisor/advisee relationship discourages providing the special attention and support that may be necessary to validate the existence of some female graduate students.

Getting the Support You Need

Recognizing that the difficulties you are experiencing are real can help. Friends, colleagues, mentors, and good role models can make it much easier. As you interact more closely with faculty, single out people you believe will give you the critical feedback and personal support you need.

Finding a good advisor is not simple, but it can be done, even if the perfect match never happens. Even if your advisor is less than ideal, many women claim that they have been able to put together a support system with the individuals who were available to them. The important thing is to realize that you may have to construct an environment that reinforces your self-image as a professional. It's not a sign of weakness to need a supportive environment.

In addition, keep in touch with people in your immediate academic community, such as other faculty members, peers, or students who can help you if your advisor doesn't. Successful students report that outside resources such as female faculty members in other departments, academic support groups, religious organizations, or even e-mail communities can provide meaningful support and critique of your work.

A Good Advisor Should:

- Help to direct and involve you in your areas of interest.

- Introduce you to the department and the scientific community, and "plug" your work.

- Counsel you and direct your research as candidly as possible, remaining sensitive to your interests and intellectual opinions.

- Assist in your course selection.

- Guide and protect you from troublesome people in the department and help you to understand and learn to negotiate the politics of your department and school.

- Offer encouragement as you "learn the ropes" and attempt new things.

- Help you to understand and meet deadlines within the department.

- Provide opportunities for you to attend and/or participate in professional meetings to become part of the academy.

- Ensure that you publish your work in appropriate archival journals.

The Importance of Having a Mentor

As one seasoned engineer observed, "Successful people have mentors; they don't just get there on their own."

Mentoring can take many forms and goes by many definitions, but generally it means having a senior, experienced person in your field who is willing and able to provide advice, wisdom, and support as you are finding your way in your chosen discipline. Mentoring can be on an arranged or a formal basis (such as being matched up with a mentor through your department, an honor society, or a mentoring group), an informal basis (having coffee with one of your former professors once a month), or even on an electronic basis such as through MentorNet. (See Web Resources at the back of the book for more information.) It can be on a daily basis or on an as-needed basis. One's mentor may be sort of a surrogate parent, someone who takes you under the wing and guides you and nurtures your development. Or it may be a faculty member who simply continues to teach you outside of the classroom, offering up knowledge about things you'll need to know in order to be successful in your field.

But realize that what you need from your mentor may not be what he or she is willing or able to offer. You may want your mentor to help you with the politics of your department, to give you emotional support when you're having writer's block,

---

HOW TO BE A MENTEE

"We all have many mentors. However, there are times when a more formal mentoring relationship is really useful. It's important in a formal mentoring relationship that there be some training and understanding about what that means for both parties. But once that relationship is established, it has to be the mentee's job to drive the relationship. A mentee can't just sit back and say, 'So what are you going to do for me?' Only the mentee knows what she wants to get out of the relationship.

"It would be foolish to assume that all pairings are going to work. It may be that you can continue in a limited fashion with very explicit goals. It may be, for example, that you can't connect with someone on a personal level; their personal life and perspective on personal lives just don't match yours, and you can't learn anything from that. But it may be you can learn something on the professional level, so it's possible that if you limit your scope it can still be productive. If it's not going to be productive, then ending it probably takes a burden off your shoulders.

"If it's not productive then it's a waste of time on both sides, and both sides should understand that. It's perfectly appropriate to say why you want to end the relationship, but you don't have to involve the gory details of why you don't want it. It is okay to say, 'This just isn't working out for me,' and make a clean end of it."

Anita Borg
Computer scientist at Xerox and president of the Institute for Women and Technology (Ph.D., Computer Science, New York University)

Read about "How to Be a Mentor" in Chapter 4.

---

or to help you find a job—but he or she might not have the resources, knowledge, or desire to help you with all of your needs. Or the reverse might be true: your mentor wants to help you more than you want to be helped. Because of all of these factors, you will need more than one mentor to draw from or fall back on. No one person can provide everything you'll need in finding your way.

Find out if your department or university offers support programs for women or graduate students. Today, many universities offer such programs that help students connect to alumni or other mentors.

REQUIREMENTS AND STANDARDS

- The Big Problem: Lack of Uniform Requirements
- How This Problem Affects Women
- Obtaining the Policies and Guidelines
- Knowing What You Need to Know

The Big Problem: Lack of Uniform Requirements

One problem you may run up against in graduate school is a lack of uniform requirements and standards for doctoral students. Each college and department within the university may have different rules and requirements concerning its graduate students—everything from admission to funding to degree completion. In some cases, standards and requirements may even vary among students within a program or department.

Academic departments may resist adopting and recording uniform standards for their Ph.D. students for a number of reasons. Among them are the fear of limiting the freedom of students in designing their own academic program, difficulty in agreeing on what those standards will be, or simple disorganization. Not having standard requirements can make life difficult for graduate students. If written requirements are available, familiarize yourself with them to avoid any potential problems. If a problem or dispute should occur during your program, you will know where to turn. Never rely solely on department folklore or on what other graduate students have told you; particularly avoid talking only to first- or second-year students. Read the written guidelines, and then merge what you learn there with what you have heard.

Some students have been hurt operating on faith alone. There are any number of stories about graduate students who have not received their degrees when expected, who have lost funding, who have had to switch projects and advisors—all because they didn't read their department's standard requirements.

Most departments have a graduate student coordinator or director of the graduate program. Ask this person for the written guidelines, but then keep your ears open for the unwritten ones.

TIP

Requirements and standards may not always be set in stone. Says one advisor of engineering graduate students, "It has been my experience that almost any requirement can be waived if there is support from the faculty."

How This Problem Affects Women

Being unfamiliar with the standard written requirements adversely affects all graduate students. However, it can have a more devastating effect on women. Like it or not, women may still be discriminated against in some academic departments; some professors and advisors may consciously or unconsciously make requirements for satisfactory performance tougher because they believe women do not belong in the academy. Consequently, the affected woman would have no proof or recourse if she was not familiar with her department's requirements.

If the requirements for completing the degree are passed down from person to person, advisor to student, a woman entering into a predominantly male arena may not be told the requirements and may be hesitant to ask what they are. A woman

Cinda-Sue Davis (Photo credit: Bob Kalmbach, University of Michigan Photo Services)

SOCIALIZING IS CRUCIAL

"I think it always helps for new graduate students to ask senior graduate students, 'What do you know now that you wish you knew when you started?'

"I think it's also important to attend social events for grad students, not only for your own sanity (which is probably the most important reason) but also because a lot of accessing unwritten rules happens in these social events. Oftentimes you think, 'Oh, I really don't have time to do that, maybe I'll skip it this time.' But it's always amazing what you can find out. So those departmental picnics, graduate welcoming events, and when faculty say, 'Please come and see me at any time'—you should do that!"

Cinda-Sue Davis
Director of the Women in Science and Engineering Program,
University of Michigan
(Ph.D., Biochemistry, University of Michigan)

may find that she is not part of the same social and recreational environment as her male colleagues. The men in the department may play tennis or racquetball with male professors, other students, or advisors in the department and, through that interaction, learn more about "the system." Good data can level the playing field.

Obtaining the Policies and Guidelines

Policies may be found in a series of materials, not just one neat booklet. You may get materials from your program, the graduate dean, or the university. Some policies may be included in your offer letter or in an annual letter you receive from your department. Ask for a department handbook, the URL for the website of policies, and the university policies that apply to graduate students. Print out and keep the policies from the website that pertain to you. If policies change in the future, you may be grandfathered in and may adhere to the old policies and not the new ones (if you choose). For example, when you began your program, a certain number of courses may have been required. If the number increases while you are still in

STANDARD COLLEGE/DEPARTMENT POLICIES
FOR GRADUATE STUDENTS

Degree Attainment

Achievement, timeline, and format requirements for the following:

- Courses and grades
- Qualifying examinations and procedures (or equivalent)
- Thesis/dissertation proposal
- Other public presentations/examinations (or equivalent)
- Graduation and degree completion specifications and a written summary of requirements
- Teaching requirements, if any
- Language proficiency requirements, if any

Financial Support

- Explicit information about requirements for awarding and the continuation of funding
- Definition of the work required for support—grading, teaching, lecturing, researching, and number of hours expected
- Information on tax implications of financial support
- Defined procedure for changing financial support, including written notification and advance warning
- Information on recourse if funding is lost or reduced

Support Services for the Student

- Advising
  1. the definition of:
     (a) the role of the advisor
     (b) the role of the student
  2. how and when the advisor is selected
  3. how to change advisors
  4. assurance that the advising process is monitored
  5. assurance that students are treated equitably
  6. procedure for written notification of "inadequate progress toward the degree"
- Orientation and other professional development courses
- Policies on:
  1. courses taken outside the department and how these will be graded
  2. when a student is defined as "all but dissertation" (ABD)
  3. residency requirements
  4. transfer from master's to doctoral programs
  5. intellectual property as it applies to graduate student and faculty collaboration

- Student rights
  1. the establishment of a form of review/redress for academic conflicts
  2. a "grandfather" policy that assures students' graduation under policies in effect at the time of matriculation or the choice to change if/as new policies are instituted

Policies and information on:

- outside employment
- the availability of summer employment
- outside fellowships
- attending conferences and seminars

Other Services

Statements on the availability of university benefits for:

- health insurance and access to health care
- parking
- emergency loans
- housing
- international student services
- teaching support and training
- professional development and support
- child care

your program, you will most likely be exempt from taking those additional courses. Having a dated hard copy of the policies will be very useful. Since policies are always changing, people forget what the policies used to be.

Knowing What You Need to Know

The boxed list above are requirements and standards you should know about before you begin any degree program. Remember, it is your responsibility to find out what the requirements and policies are. Ask your advisor, the graduate coordinator, experienced graduate students, even the department head if necessary.

If requirements are missing, or if in a particular case no one in your department knows how to proceed, speak to the chair or assistant chair of your department.

FUNDING

- The Funding Question
- Where Does the Money Come From?
- The Advisor Connection

- The Funding Package
- External Money: Fellowships
- Taxes

The Funding Question

Whether you're just beginning your graduate career or whether you've already been in a Ph.D. program for a couple of years, the question of paying for your education is, no doubt, extremely important to you. You've probably heard from faculty and peers that no one in science and engineering has to pay for graduate school. In some ways, this statement is true. For students pursuing Ph.D.s, getting a fellowship is usually possible. Master's students may have a harder time.

Paying for graduate school is different from paying for college. More than likely, your undergraduate education was either financed by your parents or paid for through a combination of scholarships, student loans, and personal and parental contributions. As a graduate student in science or engineering, your education will probably be financed by your department and academic institution.

Most graduate programs in science and engineering provide funding for their doctoral students that allows them to spend their time studying, teaching, and doing research, not scrambling for money to pay for their education. The system is generally designed to help students. But, as a prospective graduate student, it is still a good idea for you to spend time comparing schools' funding packages. If one school cannot provide you with adequate support or give you firm guarantees, think carefully before you make any permanent decisions.

Where Does the Money Come From?

At major research universities, the money to support doctoral students often comes from federal grants and contracts. Money is also granted to major universities by large national foundations, private organizations and societies, and public companies to carry out research in their best interests. Unlike most other disciplines, science and engineering receive a great deal of private and public support.

You can also be paid for working as a teaching assistant. You will want to look into the possibility of teaching at least one year, particularly if you are considering an academic career.

The Advisor Connection

In many engineering and science programs, your advisor is responsible for funding your education as early as the first year. Faculty members in science and engineering departments generally receive grants and other types of funding for their own independent research projects. You may have been accepted into a program because a particular faculty member—who will become your major advisor—

with research interests similar to yours has funding available and will be able to support you throughout your Ph.D. "Support" means that the advisor has enough funding to cover the cost of your tuition, your salary as a research assistant, and any overhead you may incur, such as the cost of office space, supplies, and a computer. (This can add up to a considerable sum. Each Ph.D. student can cost an advisor $45,000 a year at a private university.)

If your department uses this method of funding, a word to the wise: The funding your advisor has may not last forever, and there are no guarantees that she or he has to fund you for the duration of your Ph.D. program. Some students have had to switch advisors and projects in the middle of their program—even in the middle of their dissertation research—because their advisors ran out of money.

Not every department uses this method for funding doctoral students, but it is very common. Some departments pool their funding resources and use them to support all graduate students. This way, you are in less danger of losing your funding midway through the program. Moreover, your research interests may be less likely to be hampered by being forced to choose an advisor on the basis of "who has the money." One physics professor advises making sure there are at least three people you would consider working for—just in case.

Before you accept a position in graduate school, check very carefully into your department's funding system and into external funding. Some graduate departments in science and engineering will guarantee funding for three to five years. Too many students have spent extra years in graduate school, frustrated and angry, because of funding problems.

Having the option of working as a teaching assistant can provide the freedom to choose an area of study that doesn't have steady funding and can be particularly useful for theorists. While being a teaching assistant is advantageous to your long-term plans if you will be entering academia, remember that teaching can take up a good deal of your time.

The Funding Package

Doctoral education usually includes tuition remission through your department. This means that your department, or in some cases the graduate school, agrees to pay the cost of your tuition. As any longtime student can attest, tuition is only half of the battle. To help with living expenses and other costs, your academic institution or department may offer an assistantship, a grant, or an award that pays you a stipend or a regular salary of a prearranged amount.

In exchange for an assistantship, you are generally expected to work, perform research, or teach. As a research assistant, you will work for a faculty member on a particular project or experiment. A teaching assistant usually leads introductory undergraduate courses, teaches, or supervises in the laboratory and/or administers examinations and grades papers.

Graduate assistantships are awarded on the basis of scholastic merit, GRE scores, and faculty recommendations. In most Ph.D. programs in science and engineering, nearly every full-time doctoral student in a particular department receives some form of assistantship.

Mark Kantrowitz, publisher of the on-line *FinAid: The SmartStudent Guide to Financial Aid* warns prospective graduate students to ask before accepting admission about the number of hours they will be expected to spend working or teaching as part of the assistantship requirement. Programs vary widely. If you are required to spend 30 or more hours a week working, it may interfere with your classes and research. And it may be difficult to finish your program on time.

In some engineering departments, Ph.D. students receive research assistantships as part of their funding package and are only required to teach one semester as a requirement for their degree. Some students are offered teaching assistantships or a choice between teaching and research assistantships. Teaching assistant stipends often pay a comparable amount to a research assistant's stipend but require that the student take on teaching responsibility—from a lab to a section to grading. If you have a choice between teaching and research, most graduate students recommend research, particularly if you are being paid to do dissertation-related research. They suggest that if you are sure you want to teach, you can always offer to work a semester or two as a teaching assistant (the first semester to try it out, the second to work on your teaching skills). But for some people, the time in the classroom provides the emotional support needed to keep going. Knowing yourself and your needs and paying attention to them is always a good approach. On the other hand, you need to be in tune to the culture of your department: some value time spent on research over time spent teaching, and your choice of teaching over doing research may be viewed unfavorably.

External Money: Fellowships

A fellowship is a grant or an award that is similar to an undergraduate scholarship. Fellowships are referred to as "outside funding" because they come from a group or organization outside of your academic institution. When you enter a program, you may bring a fellowship award with you as part of your funding package. You are not usually expected to perform any university-assigned research or other work in exchange for the award. Fellowships can be used to cover the cost of tuition, living expenses, books, or supplies.

To receive a fellowship, you must be nominated or apply directly to the sponsoring foundation or organization. Most granting institutions conduct national searches, so expect a great deal of competition. Fellowships are awarded on the basis of scholastic merit and other considerations, including GRE scores and personal recommendations. There are several organizations and foundations that support women, and specifically women in science and engineering. Your school's fellowship office should be able to point you to on-line and hard copy resources.

Many graduate students who find outside funding report that they have a calmer experience in graduate school. Outside funding can relieve you from the pressure of proving your worthiness each year for university or departmental funding. And few advisors would turn down a "free" student.

Still, outside funding is not without its costs, particularly if it doesn't last through the whole program. Students have reported that once the fellowship is over no one feels responsible for their funding; and since no one has been looking out for them financially, no one may have been looking out for them academically. Consider that a prestigious three-year fellowship may not be as good as a not-so-prestigious multiyear one. Also be aware that with outside funding, you may not have access to the lab equipment in your program or be able to work on research teams that receive university funding. As a rule, pay careful attention to your choices.

Taxes

According to current laws, graduate student degree candidates' stipends used for tuition, fees, and school-related expenses such as books, supplies, and equipment are not subject to federal income taxes. However, money paid for room, board, and travel *is* subject to federal income tax, and the part of your stipend that represents payment for work such as teaching, research, or other services is also taxable. Tax laws are constantly evolving, and yours may be a unique situation; be sure to check with your school's financial aid office for the most current tax information.

FUNDING HEAVEN, FUNDING HELL

"My department is pretty well funded, since we get a lot of money from industry, so I had no problem getting funding. In my department, for the majority of students, your admission and your funding are tied to a certain professor; you aren't usually accepted unless there is funding for you. Sometimes the department accepts students who would be an asset to the department, but where there is no perfect fit with a professor who has funding. The department offers those students a teaching assistantship for a year, to give them time to figure out who they would want to work with and what they would want to do.

"The particulars of my funding were all detailed in my admission letter. Then, once I came here there was a fellowship that my advisor wanted me to apply for and I got it. Now I get most of my funding through the fellowship.

"I realize that I'm lucky to have the funding I have at my current school. At my previous school, where I got my bachelor's and master's degrees, I heard that they accepted more doctoral students than they had funding for, and had to force some students to finish their degrees early before their funding ran out. People left in a real hurry, and just threw stuff together at the last minute. I guess it's good in a way, because they got out earlier, but it wasn't a very relaxed atmosphere."

Gwen
doctoral student, environmental engineering

THE QUALIFIERS

- What Are Qualifiers?
- When to Take Them
- How to Prepare
- Choosing Your Committee
- Getting Through the Exam
- Keeping Your Fears in Check
- What If You Don't Pass?
- The Changing Nature of Qualifying Exams

What Are Qualifiers?

Before you begin actual work on your dissertation, you must pass what are commonly known as qualifiers or comprehensive exams. Qualifying exams are given by your department to test your breadth and depth of knowledge and to see if you have "what it takes" to earn a doctorate. These written and/or oral exams test your ability to reason, to solve problems, and to create original knowledge.

Although academic departments differ in their focus and approach to the qualifiers, the fundamental purpose of this exam is to determine whether you know your field well enough to do original research and create new knowledge.

The test is often administered by a committee of faculty members from your program. Most likely, faculty members from each area of specialization within your specific program will serve on your committee.

When to Take Them

Typically, the qualifying exams must be taken before you actually become a Ph.D. candidate. Most departments require that you complete at least one full-time semester before you take the test for the first time.

If you began your Ph.D. program with just a bachelor's degree, you will probably take the qualifiers early in your third year of study. If you already have a master's degree, plan on taking the qualifiers early in your second year.

How to Prepare

Often a department is trying to see if a student can synthesize information—that is, combine knowledge from more than one area. Ideally, your department will provide enough information on what the qualifying exams cover. However, since this may not be the case, the best way to determine what the qualifiers will test is to ask your advisor and especially other graduate students who have already completed their exam.

Other people within your department are also well versed in the qualifying exam process and probably have good advice, such as the graduate coordinator in your department, your department head or associate head, or other faculty members.

Make sure, however, that you know what will occur in your exam. The best way to face a qualifier is to understand the criteria against which your committee will judge you. This means that you need to know how your committee will operate and determine the level, format, and nature of the questions they will ask.

If you are taking qualifying exams that cover the fundamentals of your discipline, assume that you will need at least minimal competence in all areas unless you are explicitly told otherwise. You want to be able to demonstrate that you understand the prevailing theories in your field and that you understand how and why particular research methods are used.

Don't be afraid that there is too much to learn and that you won't have enough time to prepare. In the case of qualifying exams, recognizing what you don't know is crucial. You will probably be asked questions that you cannot answer. But if you can discuss how you would proceed to answer the question and why, you can demonstrate that you have some understanding of your field and are capable of using its methods to generate solutions or plans of action.

GATHER WISDOM AND PREPARE

"My qualifier includes written exams based on seven core courses and an oral. We have two written exams, one on the core and one on the specialty courses. Then we have an oral exam based on what we should have gleaned from our undergraduate materials education.

"The oral doesn't bother me. I have done enough oral presentations that I am comfortable standing up in front of people and talking, although I do know that some of the professors have a reputation for twisting the knife.

"My preparation is based closely on what more experienced graduate students have told me: talk to your advisor and to other graduate students, and prepare as much as possible.

"I talk to everybody in the research group because that's how you find out everything about the faculty, everything about what's going into the qualifier, and what it has been like for other grad students. I never discount the experience of somebody who has been there before."

Maria
doctoral student, chemistry

Choosing Your Committee

In some departments, you have the option of choosing your own qualifying exam committee. In some cases, this choice becomes a challenge because finding a committee that will benefit you both academically and professionally may be difficult.

Structuring your committee really depends on how you will be tested. Ask other graduate students in your department how they chose their committee members and what kind of experience they had with each individual member. You should always discuss potential choices with your advisor. If you can, choose committee members who can help you intellectually but also support you emotionally. Be careful not to put two people on your committee who disagree with each other on all major issues. Choose committee members who will complement each other and create a balance of different strengths and weaknesses.

Remember that you will eventually need three references when starting your career. Your advisor is one, and the other two generally come from your qualifying exam committee.

Whomever you choose, make sure to meet with your committee members as often as possible. The conferences will build up your credibility and confidence and will also give you the opportunity to ask questions about your exam material.

Getting Through the Exam

Expect your committee members to ask a number of questions concerning their field. You will probably know more about some topics than others, and you may also know more about a particular topic than your committee members do. This is perfectly natural. If you do not have any difficulty answering your committee's initial questions, be prepared for them to up the ante. Remember, the purpose of this exam is to test the limit of your knowledge and find out exactly what you know and how you think.

If your exams are oral, don't be pressured into giving immediate answers. Take whatever time you need, and use the chalkboard or paper to work out your answers. When you do answer, sound professional. As someone once said, "Presentation is everything!" Clearly state your answers and be ready to defend them.

Try not to worry if you make a mistake. If you can loop back and identify it, great, but try not to dwell on what's already happened. If you can't answer a specific question, you might tell the committee how you would derive the answer. Remember, your committee will be evaluating your overall performance, not just individual answers.

Keeping Your Fears in Check

This entire experience will test your self-confidence and ability to complete your degree. Don't dwell on the outcome of the exam, but instead focus your energy

---

HOW WELL DO YOU REASON?

"In most cases the committee is not only looking for the limits of your knowledge, but also wants to see how you reason through when you don't know something. They want to see that you know you don't know, and how you deal with that. You should be able to demonstrate what related things you know to be true, and how you would reason (and maybe even experiment) from there. Often at this point in an exam someone on the committee will actually give you information you didn't have when you came in, just to see how you would progress from a new point dynamically."

Shayna
doctoral student, mechanical engineering

---

on doing your best. You've made it into the program, and you have the skills necessary to successfully take the test.

Do a little bit of studying each day for six months prior to the exam. If the exam is oral, have someone who has taken the exam test your ability to answer tough questions on the spot.

Go into the exam with a good attitude—don't try to minimize your concern, but don't operate on the assumption that you probably will fail. Think constructively and keep in mind that in graduate school you are taught to be hypercritical, which means you may be too critical of yourself. Even though the committee is going to test your knowledge, remember that this is not an adversarial relationship. A department invests time, energy, and financial resources in graduate students; they have a legitimate interest in your success.

One way to temper your own periods of self-doubt, anger, or fear may be to talk to others who have gone through the process. Just hearing others tell you that they experienced the same doubts and fears is reassuring, especially when you talk to other women who have passed. Chances are they will know what to say to help you gain perspective.

TIP

Form a qualifiers (orals) study group with other students who are preparing. Practice answering questions with each other, regardless of subfields.

What If You Don't Pass?

It is entirely possible that you will not pass your first qualifying examinations. One engineering professor had an undecided committee on his first set of quals. He took them a second time and failed, passing finally on his third try a year later.

Many successful faculty members have similar stories. Of course, no one wants to face her qualifiers more than once, but the point is that no matter what happens, it certainly does not have to be the end of your academic career.

If you don't pass, make sure that you ask and clearly understand why. Sometimes, the reason is as simple as not having a thorough knowledge and understanding of your particular field, or your committee sees potential for success in you but decides that you're not quite ready for intensified study. In other cases, your committee may be divided on a number of issues in your work and may ask that you go back and prepare more thoroughly. Some prospective Ph.D.s have bungled their qualifiers in spite of having superior grades in preliminary course work. In a few cases, qualifying committees have decided that their students were not qualified to earn a Ph.D.

A negative decision can be especially difficult to understand in departments where there are no written criteria for passing the qualifiers. If you are involved in such a program, make sure to ask each of your committee members why you didn't pass—and press for specifics. If you feel that you've been judged unfairly, you might want to request a new committee and take the test again.

Whatever the case, try to evaluate your own performance objectively. Perhaps you really need to spend more time reading and studying. Or you may have focused too broadly: try to narrow down to what your committee specifically requires in order to pass the exam. A few of you may find yourselves rethinking the decision to pursue a Ph.D. Never give up too easily, but do spend some time with a knowledgeable person you trust examining your motives and reasons for being in a doctoral program.

The Changing Nature of Qualifying Exams

Always make sure your information on qualifiers is updated—the qualification system does not just exist, it evolves. It has changed and continues to change for a variety of reasons, including changing department standards and power structure, frustration with the results of the past system, and even at the request of (and because of complaints from) graduate students.

THE DISSERTATION

- What Is the Dissertation?
- Choosing a Topic
- Choosing a Committee
- Writing
- The Defense

For many graduate students, writing the dissertation is the quintessential love/hate experience. On one hand, as a student who has spent the last several years of your life toiling away, researching, learning, and studying, tackling the monster of the dissertation should be a welcome challenge. It can be exciting to put the results of your own original research down on paper. On the other hand, being asked to produce what could be the biggest project of your life to date can seem insurmountable. No one said it was going to be easy!

What Is the Dissertation?

A doctoral dissertation is an original piece of work that contributes to the theory and practice of your field of study. The ideas for your dissertation are generated from questions that have not yet been adequately answered. Your qualifying examinations were designed to test your mastery of the existing knowledge in your chosen field. Writing the dissertation should take you one step further: you will examine the unexplained in your discipline and draw conclusions from what you discover. Writing a dissertation identifies you as an "expert" in that field because no one has traveled down the same road before.

Choosing a Topic

In some graduate programs, defining a topic is part of the qualifying process. In most programs, it is part of your advisor's research. In others, you have complete freedom and responsibility to choose a dissertation topic on your own. To begin the process, you must first choose a topic that interests you. You will be spending most of your time in the last six to twelve months or so of graduate school thinking, writing, and stewing about this topic, so you should be sufficiently excited about it to sustain an interest. Your topic should involve a problem or an unanswered question related to your field of study and not be too broad. You need to make sure that you can answer the question in a reasonable amount of time (and number of pages). Talk to your advisor about this, however; there are big variations from department to department, advisor to advisor. Expect 100 pages but usually not more than 300.

Work with your advisor when choosing a dissertation topic. She or he can help you by defining the expectations of your department and may have several specific suggestions. Look over the written dissertation guidelines for your department, if available, and look at other dissertations produced by students in your department. Try to talk to these people individually for their help and advice. You should also discuss your ideas with your fellow graduate students for suggestions and constructive criticism.

To start thinking about potential topics, read whatever you can about what interests you in your field of study. Perhaps you will find topics that have been overlooked or not fully explained. Read related dissertations and other projects. Attend seminars, colloquia, and presentations involving experts and other students studying

---

LOVE YOUR TOPIC

"You definitely have to pick a dissertation topic that is a personal passion, otherwise you're going to be sick of it by the time you finish. If you didn't start out in something that you're really interested in, you're going to regret the fact that you went through the process. Not everyone is as free to do that as they would like to be, but you can't let yourself be pushed too far by the needs of your research group or your advisor's personal opinion. It's got to be your work; that's the whole point of doing it. And it's got to be something you really love."

Lauren
postdoctoral researcher, computer science

---

similar disciplines. Absorb as much information as you can. You never know when something will jump out at you and spark your interest.

If you're stuck or having a hard time, consider taking additional classes or performing new or different research. Or try to research and study in more detail. You may come across something through one of these experiences that will help you narrow down your interests.

Finally, before making a final choice, consider the following:

- Does your topic interest you? Will it sufficiently challenge and excite you for the next year or two of your life?

- Are you comfortable being identified as a specialist in the field? Are you willing to be defined as the expert on the subject?

- Does your topic make a contribution to your field? (Your topic does not have to be the definitive piece of scholarship in your area, but it should bring to light something that was previously unknown.)

- Is your topic original? Are you willing to master the skill of independent thinking?

- Is your topic reasonable? Is it well-defined and sufficiently limited in scope? Can you finish it in the appropriate length of time?

Choosing a Committee

Your dissertation committee typically consists of at least three to four faculty members from your field who will evaluate your progress and final dissertation. They help you shape the body of your topic, research, and dissertation by talking with you, by guiding your research, and by reading and critiquing drafts. When you have finished writing and revising the dissertation, your committee hears you de-

Sara Majetich (Photo credit: Ken Andreyo)

STAY INTERESTED BUT FLEXIBLE

"Your topic usually changes somewhat as you do the research. Make sure you're interested, that in doing the research you will develop the skills you want and that it will send you along the career path you want. Apart from that, be flexible."

Sara Majetich
Physics professor, Carnegie Mellon University
(Ph.D., Physics, University of Georgia)

fend your research and findings in a formal presentation to the academic community.

Your thesis advisor may be the one who typically chooses the committee or will at least help you choose it. If you do have a hand in choosing your committee, many graduate students suggest that you use the same criteria you employed when selecting an advisor. After all, these individuals will have a significant impact on your future reputation as a researcher and scholar. They advise that the ideal committee should have one faculty member representing each area of expertise related to your topic and should include a "big name" in the field to help increase your own reputation. Make sure that you can trust your committee members to be responsible. This includes reading and returning critiqued drafts in a timely fashion and providing constructive criticism, encouragement, and/or praise when appropriate. Try to find a committee that will agree on the scope and focus of your work. You want to avoid the endless cycle of "changing and changing back" whenever possible. Also try to choose a committee that will be around when you need them to be. You may want to avoid faculty who will be on leave when you are scheduled to write and present your dissertation. However, if the national expert in your discipline is willing to serve on your committee, by all means, put her on! Sometimes having the most credible advice is an even exchange for regular attention.

Whatever the case, try to remain on friendly terms with your committee during the entire dissertation writing process. Try to balance competing factions in your group, and check with each of them before making any major changes to your work.

Writing

The Prospectus.   To begin, many graduate students are required to submit a prospectus or proposal on their research project. This, too, varies from field to field. The prospectus is basically an informal argument for your research agenda that:

- outlines your area of study,
- places your topic in relation to the field,
- demonstrates how your particular questions arose out of this existing knowledge,
- defines and defends your proposed plan of research and study, and
- argues for the significance of your work.

Your prospectus is sometimes submitted to your dissertation committee as part of a preliminary oral examination. If your committee approves your plan, you are allowed to begin your formal dissertation research. If they find problems with your prospectus, such as a too broadly or narrowly defined topic, you will be asked to re-think your topic and to rewrite your proposal.

Develop a Plan.   Once you obtain approval to begin research and writing, for-mulate a plan of action. Too many students face problems due to a lack of structure and direction. Getting and staying organized are the keys to success.

TIP

Plan far ahead when ordering equipment for experiments. It may take a long time for it to arrive.

Create a plan before you start. It is definitely worth taking the time to outline how you plan to tackle the task of the dissertation. A realistic and detailed timetable for writing the proposal, conducting your research, analyzing the data, writing your thesis, and defending it can be an enormous help. Remember, your plan will proba-bly change, as will your original ideas and research, but any plan is better than none at all.

Establish a routine for working. Some students approach working toward their dissertation like working at a job—full time, every day, except perhaps one day on the weekend. Others do lab work in the morning and reserve the afternoons for tak-ing notes and writing. Whatever method works for you, stick to it. The more orga-nized you are, the less likely you will be distracted by lack of structure or feelings of being overwhelmed.

Prioritize your tasks at hand, and keep a log of accomplishments. You will find this helpful in managing your time and estimating when you will finish the entire project! Keeping a log will also help you to set realistic goals and keep you from becoming discouraged. All of these suggestions for organization are, just that—

DISCIPLINE, DISCIPLINE, DISCIPLINE

"Writing really takes discipline. Thesis avoidance is an art; some people get so good at it that they never finish. Discipline for me was definite goals of finishing definite chapters by definite points in time. And my advisor and my committee were expecting things on those days, so it had to be out and done. Also it meant giving up everything else. Everyone says for at least six months; in my case, it was more like nine or ten months. You just have to give up the rest of your life. Your family, your personal relationships, everything around you suffers for that; but everybody has to realize it's short term and it has to be done. In the end when you sit down to write the bloody thing, it has to be the only thing you're doing in your life."

Katherine
recent mathematics graduate

mere suggestions. Find what works for you, and remember to stick to it! Few people want to end up a perpetual ABD (all but dissertation) candidate.

Writing…and Writing…and Writing… Some of you may have had the foresight to write chapters of your dissertation throughout the researching process. If you have opted to complete the research and then begin the dubious task of putting the results on paper, this section is for you.

After you have conducted your research and gathered and analyzed the data, you have to put your thoughts down on paper. Writing can be the biggest challenge for many graduate students. One graduate student suggested taking notes throughout the entire dissertation process and keeping copious files. It is also wise to keep a copy of your prospectus close at hand. When you're writing, it can be easy to lose sight of why you ever chose your topic in the first place. Reading and incorporating the points of your prospectus can remind you.

Using a software program to keep track of your citations can be a time- and sanity-saver. Check out the various packages available to see what's best for you.

When writing the dissertation, remain in close contact with your advisor and committee. Other graduate students in your department may also provide useful advice and support to help you stay motivated. Some people write from a detailed outline, whereas others prefer a core dump and editing approach. Whatever you do, do not write an entire first draft before showing it to your advisor. Share a general outline, and plan and review each chapter (or several) at a time.

Expect to be overwhelmed at times, or bored, or frustrated, or angry. Taking on such an enormous project can affect even the most organized and stable person. Many of you will fear that you will never finish. Given that this may be the biggest project of your life, these feelings are perfectly understandable. Try to keep things in perspective and discuss your feelings with other graduate students. Don't give up working, and try to write at least a little each day. Keeping to a routine, as always, will help—but if you're stuck, give yourself a break. Doing a completely unrelated activity may be the impetus you need to cure writer's block.

TIPS

To overcome writer's block

- **Do something completely brainless (exercise, watch television).**
- **Trick yourself into writing. Write too much for too long, deleting nothing as you type.**
- **Speaking may be easier than writing. Give a talk (even if just to yourself), then transcribe it.**
- **You don't have to start at the beginning; start with something you feel comfortable with.**
- **Explain your topic to your significant other or a family member.**
- **Physical environment matters: Clean your desk. Change your lighting to incandescent. Go somewhere else.**
- **Sit down and resolve to stay seated until something specific happens. Do not even get up for coffee!**
- **Don't get hung up on your first draft. The greatest writers have horrible first drafts. Revise on paper, not at a computer.**

(From "Overcoming the Dissertation Blues: Concrete Ideas on How to Get Through It," a Graduate Student Seminar presented by Illah Nourbakhsh, assistant professor of Robotics, Carnegie Mellon University; inspired by Patrick H. Winston's "Breaking Writer's Cramp," unpublished work)

Every Ph.D. will tell you about the blues. It is perfectly natural to stall out in the middle of such a large project. Above all else, part of the purpose of writing a dissertation is to conclusively demonstrate your ability as a scholar. Remind yourself that your work only needs to be original and to make a contribution to your field. Quoting her advisor, one graduate student tells us that a good dissertation only needs to accomplish *one* of the following things:

- Open up a new area.
- Provide a unifying framework.
- Resolve a long-standing question.
- Thoroughly explore an area.
- Contradict existing knowledge.
- Experimentally validate a theory.
- Produce an ambitious system.
- Provide empirical data.

- Derive superior algorithms or methods.
- Develop a new methodology.
- Develop a new tool.
- Produce a negative result.

Just because one of these things has been done, however, doesn't mean that it's necessarily worth a thesis. In general, the advisor decides when enough results are in to constitute a thesis. The committee can sway the advisor a little, but what is important is to make sure enough work has been done so that the student doesn't get stuck at the defense.

Don't expect too much of yourself. Your dissertation does not have to be your magnum opus. There is a difference between getting something credible done and writing *the* quintessential treatise on your particular topic.

The Defense

The defense may be the first formal presentation of your dissertation before your committee and the rest of the academic community. Although the actual defense may be a formality, it is your chance to shine as an academic. Many people will see you and form an opinion of your research style and work. Remember that in some departments, many changes are typically made to the thesis after the defense. Make sure your committee is not seeing your first draft, however. Usually your advisor corrects the first few versions, chapter by chapter.

Before you defend your work, make sure that your committee has seen your final draft and that you have incorporated all of their changes that you are willing to make. If you leave critical changes out or ignore the advice of one member of your committee, you may be at risk during your defense. Know your material thoroughly and be prepared for any kind of question regarding your idea, research, or punctuation style. You should be prepared to answer almost anything about your thesis work.

After your defense, you may have some or a lot of rewriting to do, but when it is all done, don't forget to try to get your work published. Some of you may have already had original or coauthored articles in print, or you may have already published chapters of your dissertation during the writing process. Publishing is critical to your future success as a researcher inside or outside the academy, and it is never too early to start. Almost all universities have agreements with University Microfilms to have their students' doctoral dissertations "published" for archival purposes.

CHAPTER 3

POTENTIAL STORMS

"Self-doubt is natural for almost all who go through a Ph.D. program. Women may have more of this self-doubt, or maybe they just acknowledge it more openly.

"Despite the pressure, it is the only time in one's life that the intellectual demand is primarily and legitimately directed toward discovering and developing your own strengths. This is where one develops one's intellectual integrity. This means both grounding oneself in a core of sound, good knowledge and building a whole from this core."

Indira Nair
Vice Provost for Education and
Associate Professor of Engineering and Public Policy
Carnegie Mellon University
(Ph.D., Physics, Northwestern University)

SELF-ESTEEM

- Self-Esteem and the Graduate Student
- Relationships and Self-Esteem
- Building a Support System
- Helpful Hints

Self-Esteem and the Graduate Student

We all know that graduate school is tough. A typical doctoral candidate has a rigorous schedule of challenging classes to complete, qualifying exams to pass, a thesis to research, write, and defend, and some semblance of a life to live—when there's time. On top of all that, graduate students have to learn how to live and work in a politically charged university system, how to work on a largely independent basis, and how to deal with each other, faculty, and a department. Prospective Ph.D.s have to cultivate a working relationship with an advisor, teach and supervise other students, and carve a niche for themselves in the research community.

Women graduate students have still other waters to navigate. The male environment, the scarcity of female role models and mentors, the dearth of support and/or understanding, the closed doors and hidden agendas, and the general unwillingness to include women in some departments, faculty, and graduate schools can all weigh heavily on a female graduate student and make her experience all the more difficult. Female graduate students with self-esteem problems will have an even tougher time.

Although some self-doubt is completely normal, success in graduate school requires a certain level of confidence. In doctoral work you need to test yourself and stretch your talents to the maximum. You need to explore your interests and desires, work hard, and produce positive results. You also need to be proud of yourself and your accomplishments, and to do a terrific job. Some women with low self-esteem deeply doubt themselves and their abilities to make it through. They question whether they belong in graduate school. They think, "What am I doing here? I don't deserve to be here." They fear they aren't smart enough or good enough to proceed. Some level of low self-image is manageable; the crucial question is how much it affects performance. If you find yourself almost always negative and complaining, feeling hopeless or apathetic, lacking clear focus or goals, you obviously need help.

Women with low self-esteem may also suffer in less dramatic ways. For example, they may not be as assertive or vocal when defending their research or academic interests. Competition with other bright and talented graduate students may cause them to be hostile or feel overwhelmed. Some may have a difficult time finding an advisor, a mentor, or other graduate students to turn to for support. Too often people with low self-esteem turn inward and do not reach out for help. Imagine a woman in science or engineering with low self-esteem. In some cases, her environment is already unwelcoming enough—if she is not internally equipped to successfully deal with it and does not seek help, it is easy to fall behind and never catch up.

Relationships and Self-Esteem

Many women in science and engineering find it difficult to relate to others in their department. More than likely, there are very few women in the department, and many men may not understand the situation or be particularly supportive of women's needs. You may find that some of your male professors and colleagues don't believe or appreciate your other commitments. All of these factors can contribute to feelings of unworthiness or exacerbate such feelings that are already there.

Because most women face extra hurdles, they may need extra emotional support and reinforcement from their advisor or other faculty members and peers. They may also feel the need for more emotional and intellectual reinforcement than is traditional in graduate student/professor relationships. Although some faculty members may overtly or covertly "marginalize" such needs, if you feel any of these needs, you are not wrong. Find the support you need. There are usually many sources: other faculty members, senior graduate students, counselors, and administrators may be sources of advice and support. Ask for help.

Because women are raised in a society that continually signals that they are of less intellectual value than men, it is not surprising that some have a difficult time sustaining a view of themselves as people who create knowledge and set rules. Indeed, some women report that they feel like "impostors" who are always just getting by. Since these same cultural biases also shape the expectations and responses of faculty, some may also perceive women as less credible and knowledgeable. Even when faculty members do not share this belief, the nature of the professional advisor/advisee relationship discourages providing the special attention and support that may be necessary to validate some women as professionals.

TIP

Keep a running list or file of your accomplishments: top grades in tough courses, glowing letters of recommendation, and so on. Refer to this list if you're feeling down.

Many graduate students report that it is difficult to distinguish between the problems that all graduate students experience in a demanding context of developing oneself as a scholar and those that stem from race or gender bias. This difficulty is compounded for women in science and engineering because it is harder to find

---

GROUP SUPPORT

"While the professors were supportive, most of them were busy. Therefore, my peers turned out to be my best support group, especially the people in my lab. Despite the competitiveness of the environment, we worked together to learn and to get through the program. If someone in the group was taking a qualifying exam, everyone else would take time out to help that person study. I was the only woman in the lab group, but I wasn't uncomfortable in that situation. There were some specific activities that were for women; for example, the graduate women had an informal organization that we called GWE (Graduate Women of Engineering). There weren't many of us at the time, so it was easy for us to get together and talk about mutual interests and issues. However, most of the problems we were having the guys were having too, so there wasn't a need for any gender-related distinctions."

Carol Wilson
Member of the Technical Staff,
Hewlett Packard Laboratories
(Ph.D., Mechanical Engineering, University of California–Berkeley)

From *Journeys of Women in Science and Engineering: No Universal Constants*

role models in their immediate environment who understand their situation. They may also feel that they have no one they can confide in or talk to about feeling inadequate without reinforcing stereotypes. Graduate students tell us that if they can't share their fears or concerns with others because they are afraid this will only confirm the suspicions that women really are less capable, it makes it worse. Some avoid contact that might help them resolve their dilemma, such as asking questions or getting help with difficult material, for fear that others may see them as vulnerable or weak. Most recommend finding someone to talk to.

Building a Support System

"Dips in self-esteem are to be expected at certain times for anyone," notes one woman professor. "The freshman year of college, the first year of graduate school, just before or during thesis writing or any time you are applying for admission, a scholarship, or a job—understanding that drops in self-esteem happen at these times for most people will help you understand that you are not alone."

SEEKING HELP, FINDING FRIENDS

"After the coursework was over and I began concentrating on research, I found myself at the bottom of the lab food-chain, where the post-docs in the lab would be unpleasant to me if they had had a bad day. Many times, I would come away thinking I was at fault and my opinion of myself would slide even lower. My advisor is a very brilliant person, and often when he'd have to explain something really basic to me, I felt like I was disappointing him miserably.

"My plunging self-esteem and other adjustment problems led to my getting depressed and seeking counseling. This was one of the best things that I could have done for myself, since it helped make me aware that I didn't have to be Superwoman to be a Ph.D. student. I formed a small network of good friends who were also women in Ph.D. programs. Many a time when we'd compare notes, we'd find all of us going through the similar situations. This was reassuring, since it meant that we were all quite normal. I also found wonderful mentors and friends in a woman professor and a woman research faculty member. An important result of this was that I began talking to many of the male authority figures and actively asking for feedback and discussing aspects of my life (such as my career) that I'd have never discussed otherwise. I also realized my value to my group, the department and the community as a whole as my roles in various committees and organizations began getting acknowledged. One of the greatest slaps on the back happened when I won an award for a talk that I gave at a conference. It was a shining trophy for my victory in the war against sliding self-esteem, depression and loneliness."

Pam
doctoral student, chemistry

If you or someone you know is burdened with any of these problems, the smartest thing to do is to seek help. Whether that's talking to people you trust and respect, seeing a professional, meeting in a formal or an informal support group face-to-face or on-line, or reading current literature—recognizing the problem and doing something about it is what is most important.

When you choose help from the many people around, choose well. Carefully chosen friends and colleagues, mentors, and role models can make it so much easier to sustain a positive view of yourself. Once you better appreciate yourself, you are more capable of evaluating your behavior in relation to your own perceived model of acceptable behavior. Since graduate school is an acculturation process, you can benefit from people who can "show you the ropes" and help you to interpret the responses you get in this new "culture." One of the most immediate sources will undoubtedly be your classmates and office mates, who are struggling with you to re-frame a view of themselves as capable, intelligent players in a new environment.

As you interact more closely with faculty, your advisor, and others, you can better select people from whom you will receive critical feedback on your work and those who can provide the personal support you need.

Most students are able to put together the support system they need. The important thing is to realize that there is nothing wrong if you have to construct an environment that reinforces your self-image of yourself as a professional and a colleague. It's not a sign of weakness to need a supportive environment.

Helpful Hints

- If you have constant feelings of helplessness and frustration, there's nothing wrong with seeking professional help. Many graduate students find it enormously helpful.

- Limit your exposure to stressful situations as much as possible and seek out those people who encourage you. If you cannot avoid an individual or situation that you find stressful, find ways to shorten or limit the times and ways in which you are involved in such situations. Look for ways to balance these with the situations or individuals that help you to feel better and make progress.

- Get perspective on any feelings of inadequacy. Are they part of the normal graduate student learning curve? Are others responding to you because of their biases? If so, how can you effectively respond? Seek people who can validate your concerns, help you to analyze what is happening, and give concrete advice.

- Actively maintain interactions with people in your immediate academic community, such as faculty members, peers, or your students who can and will acknowledge you as a professional and an academic. Also use outside resources such as nondepartmental faculty members, academic support groups, or even e-mail communities to provide support and critique of your work.

- Be patient with yourself throughout this process of "redefinition." Accept the fact that you are working through new relationships and testing new behavior in a foreign environment. Don't expect to feel "successful" overnight. Expect some feelings of inadequacy; at least at first, there will probably be a lot of discrepancy between who you are and who you want to be. In addition, don't expect yourself to ever "get it all right." There will always be discrepancies, to a greater or lesser extent, between your "ideal" and your "real" states. What you are aiming for is that state of harmony in which you feel you are generally acting in ways that are consistent with who you want to be.

- Create a realistic image of your professional self. Construct an image of yourself as a successful academic that leaves room to make mistakes and be less than totally prepared in all aspects of your life. For example, if your association with the word "scholar" is a learned old gentleman settled into a reading chair in a library lined with leather-bound books (with a wife serving him tea, perhaps?), then you may have to do some major adjustments to your image of "the scholar" to accommodate yourself.

- Look at the people you consider to be good role models and realistically evaluate what it means to be "good." In other words, make a clear distinction between being a "good academic" or a "good scholar" and being "perfect." When you take the time to evaluate these people and their life choices, you may be surprised at how really "human" even the best academics are.

- Set realistic goals. Celebrate small achievements along the way to a bigger success. Clarify your vision and your values. In other words, figure out what you want and go for it! Be happy with what you can do for the moment, accept setbacks, and keep on moving upward!

FEELING ALONE

- One in a Million
- The Lone Researcher
- A Long Way from Home
- Getting Connected

One in a Million

Experiencing feelings of isolation and loneliness is perfectly natural; many women in graduate school share them. As one graduate student told us, "It comes with the territory."

In your department, you may be one of only a few women. There may be, at most, one or two women faculty members and a handful of women students; the only other women in the department are the secretaries. The faculty are undoubtedly mostly male, as are the department head and your advisor. Why? Only 9% of the total number of people pursuing doctoral degrees in science or engineering are women. Only 36% of women with doctoral degrees continue to teach in college environments. Even with so many schools in the country offering programs in science and engineering, having a balanced number of women teach in your department is unlikely. The majority of professors and students in science and engineering are men.

In an environment where you are in the minority, it is not surprising if it is difficult to make connections and to find people you can talk to. Many men may not understand your feelings because they had quite a different graduate school experience. Although you may have male friends in your program, you may still find it difficult to have things in common educationally and emotionally. Simply put, without other women—those who have experienced the same difficulties and challenges in an academic environment, and those who are most like you and with whom you have things in common—you may end up feeling isolated or alone.

Dr. Shirley Jackson was the first African American woman to earn a doctorate in any field from MIT, the first woman and the first African American to chair the Nuclear Regulatory Commission, and the first African American woman to head a top technological university (Rensselaer Polytechnic Institute). She recalls that she did experience some isolation initially as an undergraduate in physics at MIT, but once she and her classmates got to know one another, she came to be judged on her

DOING SCIENCE HER WAY

"Typically, a male graduate student will establish a set of colleagues whom he will know his whole life. I didn't really have that. The male students would pick up the phone to call each other about problems, but no one would pick up the phone to call me. I didn't work on problems that everybody else was working on—I found it very distasteful to have to compete that way. I chose problems that I thought would be of value to astronomy when I had finished.

". . . . When I was in school [in the 1950s], I was continually told to go off and find something else to study or told that I wouldn't get a job as an astronomer. But I just didn't listen. If it's something you really want to do, you just have to go do it; and maybe have the courage to do it a little differently."

Vera Rubin
Staff Member, Department of Terrestrial Magnetism,
The Carnegie Institution of Washington
(Ph.D., Astronomy, Georgetown University)

From *Journeys of Women in Science and Engineering: No Universal Constants*

merits. "There is a certain ironic advantage to standing out among others," she notes, "people tend to remember who one is" (Ambrose et al., 1997).

The Lone Researcher

Another factor that may be contributing to your loneliness is the nature of graduate work. Unlike your undergraduate education, much of your time in graduate school may be spent on your own study and research. By definition, making some area of knowledge your own, delving into it according to your perspective, requires working by yourself to a certain degree. The first year or two involves classes and study with other students. After that, you may spend much or most of your time either alone or with your advisor.

Also, most science and engineering departments are compartmentalized; the different fields of study, programs, and schools are separated within the department. Everyone is very busy, and most people only have time to talk about their particular area. It can be difficult to branch out and find companionship.

A Long Way from Home

Living in a new town away from family and friends and starting school in a completely new environment also contribute to the isolation of graduate students.

When you started graduate school, your entire life probably changed. You left your former city or home, your friends, and your entire base of support. At least initially, most of that support is no longer available. It's no wonder you feel isolated.

Also, you may be the only person in your family or circle of friends who has pursued a graduate education. Your college friends may already be working. Your parents and siblings may not have gone to college, or also started working after receiving their bachelor's degree. They may not understand your drive and desire for a graduate degree, and they may underestimate the amount of pressure you are under from day one.

TIP

When you first start graduate school, accept any and all invitations to social events. Now is when you establish your peer network, a support group you will need throughout your graduate school career.

It takes time and energy to establish new friendships—and time and energy are the two resources that are most in demand in graduate school. The temptation is to "let things slide" because you feel the immediate demands of school are more important. Up to a point, this is a healthy response to the demands of graduate school. However, if you don't allow yourself any time to socialize or sort through your own

CULTURE SHOCK

"Now that I look back on it, coming to graduate school here in the U.S. was one of the biggest challenges that I have ever taken on. I had lived in a hostel during my undergraduate education in India, about 800 miles away from home, only going home for long holidays like summer and winter breaks. Even when I was home, I'd hardly ever stay around. I didn't think coming a distance about 10 times as much would be a great issue. I would have fewer trips home—one vacation a year. But it had seemed to me at that time that it would be a great adventure to go far away from home by myself, to live in a totally new place and culture, and to meet new people.

"Little did I imagine the culture shock that would hit me within months of embarking on my new, 'exciting' life! Having to not only take care of my work but also myself, unable to make many friends (back at home I had dozens), not having the opportunity to pursue activities (like Indian dance) that I had pursued at home, having to deal with an unfamiliar and sometimes annoying culture and, most importantly, not too sure about my choice of going to graduate school. I was very confused, unhappy and frequently went into depression.

"I had not thought that undergraduate and graduate school would be so different. Grades didn't seem to matter very much anymore, and that was a huge drawback because I did not see any other means for feedback on my performance. Having to face all of this without the huge support system that I had had back in India consisting of friends, parents, and other relatives became almost impossible to contemplate and I would often find myself having panic attacks.

"I began making an effort to make friends both inside and outside my department, in some cases going to the extent of asking people to be with me when I was depressed. I also joined a lot of organizations on campus and volunteered for some committees and off-campus volunteer organizations. I needed to actively deal with my loneliness and look for outlets outside my department. In the past four years, these efforts have paid off well, and I feel less of a misfit both in graduate school and in the U.S."

Geeta
doctoral student, chemical engineering

problems, you may find yourself on a curve of diminishing returns, feeling overwhelmed and isolated.

Getting Connected

If feelings of hopelessness and loneliness settle in, you may find it difficult to see the light at the end of the tunnel—it's only human. You may begin to wonder,

"Will I ever make it through grad school? What am I doing here? Am I smart enough to earn a Ph.D.? Am I resourceful enough? How can I do it when there's no one to help me?"

Again, the answer is: there are people to help you, and there are ways to "get connected" in graduate school. Graduate school is an important test of you—of your intelligence, your confidence, and your desire for a degree. But no one said you had to do it alone. Yes, the system is sometimes unfriendly, and many feel that there are higher hurdles for women to overcome, but there is help if you seek it out. Just as with everything else, you need to make an extra effort to create an experience that is enjoyable and rewarding.

Even though developing the ability to think independently is inherently isolating, it is also inherently collaborative. Your advisors and other faculty will be the sounding board for your work, as you will be for theirs. As you develop the expertise and skills to critique and produce original work, you will become an essential member of the community of scholars.

Remember, there are many people in your school or department who are completely on your side. There are faculty members, older graduate students, and postdoctoral fellows and administrators who fully recognize the difficulties women experience in graduate school, and are willing to help.

How to Get Connected

- Recognize that the people in your department share your interests and are there to support and stimulate your intellectual growth.

- Try to meet with faculty informally. Once you have a sense of who they are as individuals, you will feel more comfortable interacting with them professionally. Also, seek out women faculty members who might be helpful whether or not they teach in your discipline. Although not every woman faculty member will see these as her issues, most women faculty will be supportive. After all, they were once in your shoes.

- Be aware that some universities sponsor luncheons, dinners, and other gatherings that allow women graduate students and faculty to get together—to participate in structured discussions, mingle informally, or do some serious networking.

- If you have a problem with self-esteem, or think you do, try to get some help. Help can come in the form of meeting with a group or individual counselor, reading books, listening to self-help tapes, or keeping a "savings account" of positive experiences (papers with good grades, letters of recommendation, positive teaching evaluations) to look over when the going gets rough.

- Participate in student organization activities. Many institutions have an active graduate student organization that sponsors social activities, speakers, and programs throughout the academic year and serves as an advocate for graduate student concerns to the university faculty and administration. Through these student organizations you can meet other graduate students, share your questions and concerns, and involve yourself in graduate student affairs. Get involved!

- Attend conferences that will allow you to stay on the cutting edge of what's happening in your field, and to meet others who are as passionate about your interests as you are. While learning more, you will also be getting to know more people—both faculty from other universities and other graduate students with whom you may have much in common. (Conferences sponsored by national organizations frequently hold break-out sessions just for grad students.)

- Look into electronic networks such as Systers, specialized societies for your field, organizations that are dedicated to promoting equity for women, and mentoring groups such as MentorNet. (See Web Resources for a list of groups.)

- Keep your goal in perspective. You are here to earn a degree and begin your career. Being a woman in the sciences or engineering isn't easy—but it probably wasn't easy in undergraduate school either.

LEARNING BY CRITIQUE

- Why We Hate It
- What You Can Do
- Learning How *Not* to Be a "Nice" Girl

As an undergraduate, the method of learning was probably something like this: your professor lectured on a series of topics on which you were later tested. Whether there was little or a lot of discussion about the material, the lesson was clearly defined. Your professor had certain objectives she or he wanted to cover, and the material was chosen to illustrate definitive points. You were learning existing knowledge.

In graduate school, the method of learning is very different—and difficult for a number of women. At first, your classes will seem like those in college—you'll study and learn an existing body of knowledge. However, the ultimate goal of attending these classes is to help you search out questions and define your research interests—not to specifically "learn" a lesson.

As you progress in your graduate work, most of your learning will come through a series of formal and informal exchanges in which others, both faculty and peers, will challenge and test your ideas. As your research progresses, you will be expected to share and discuss your findings with others. If you haven't already encountered questioning of this sort, imagine how novice lawyers are trained. We've all seen and heard about law professors "grilling" their students on case law. Although the questioning may be intense, the professor is really trying to test the student's analytical, reasoning, and communication skills.

Why We Hate It

Many women perceive insistent questioning as harsh and negative, or as a personal attack. They may feel particularly uncomfortable with situations in which their understanding is continually challenged. Women may feel vulnerable as a result of stereotypes portraying them as "dumb," or they may lack the confidence and self-esteem necessary to handle intense inquiry. By asking questions and continually challenging their reasoning, many women feel that a professor is commenting on their intelligence or worthiness as graduate students. Some report feeling strange or unworthy for asking "too many questions" or for presenting new ideas.

A SURVIVAL STRATEGY

"For some reason, women put themselves completely, utterly and wholly into their work. And that may be because traditionally our work has been other human beings, and that's necessary if you're going to do a good job working with other human beings.

"Men are often shocked at how personally women take criticism, and the only thing I can say is that until there are more women working in nontraditional fields, we have to learn to play by those rules as best we can because we're never going to make it if we don't. You can't change the fact that you put your heart and soul into the work and that it is personal, but you can adapt, adjust, and develop mental strategies for understanding that something might not be a personal criticism even though it's a very directed critique of your work.

"Negative feedback doesn't always mean that you're wrong; it means that the person disagrees or is trying to make your work better. When someone says, 'Here's a place where I don't think your work is very good; here's how you can make it better,' it's not the same as 'You're bad, you did it wrong.' The mental strategy has to be to see how they see it, and try to react to that. That's the only thing I've been able to do to survive, and it works pretty well. That's what the guys do. I'm not satisfied with it in the long run, but it's a survival strategy."

Ann
doctoral student, biology

Although some women can positively respond to learning through critique, many internalize the criticism. Conversely, professors may believe that they are pushing a student to explore new areas and to think independently. However, a female student lacking in confidence may only hear, "you're wrong and you don't belong here."

The nature of learning through critique—which is at the core of the graduate school experience—lends itself to another problem that is closely related to internalizing criticism. Although you are learning to test and evaluate your own ideas, there are very few rules to guide your progress. An idea is not perfect the first time around—learning to be an original thinker takes a lot of trial and error. Once you begin to understand the nature of original research, you'll come to understand the tenuous nature of knowledge. Many "right answers" change over time.

What You Can Do

Self-esteem and socialization are at the root of many women's difficulty in dealing with new methods of learning in graduate school. Although problems like these are not going to be solved overnight, there are steps you can take to make your experience more rewarding and satisfying.

Hints on Feeling Confident

- Set realistic goals. You are not going to know everything the minute you start graduate school. Why put that kind of pressure on yourself? Remember, you're here to learn.

- Recognize that many things can only be learned through trial and error. You will make mistakes, you will be embarrassed once in a while—it happens to everyone! When you make mistakes, focus on what you can and did learn from it. If at first you don't succeed. . . .

- Talk to others about your experience. Older and more experienced graduate students, faculty members, counselors—they all understand what you're going through, and many of them have already been there.

- Realize that you can never be completely prepared for everything. You may think because you're a graduate student in a prestigious university, you should automatically know what's going on at all times. There is a tremendous amount of pressure on women, especially women in science and engineering, to prove themselves. Go ahead with your work, and if you find you're lacking in a certain area, do what you need to improve. Don't judge yourself too harshly.

- Ask for what you want. Your professors are there to help you—they aren't out to get you, no matter how it may seem. Realize, however, that some of your superiors may be oblivious to your concerns and may forget how it feels to be critically evaluated.

- Negative feedback does not always mean that you are wrong. Try to evaluate the comments on your work objectively, and then make a rational decision about whether or not to open a discussion or a debate. This is important. Some criticism is wrong; in general, criticism is usually opinion. Learn to evaluate criticism (opinion) and decide if it's valued. Just because it was said by a faculty member doesn't make it right! Women are more likely than men

---

FINDING YOUR PLACE

"There are really two things that you need to make it. One is to be able to ask the right question, not necessarily know the answer, but ask the questions, to see right away where the information is going. (I let data speak to me.) And the other is to be accepted by one's peers.

"I remember sitting in the student lounge and some people from my class were arguing, 'I don't understand this.' And I just stepped up and said, 'Well, it's right here—in the notes.'

"And then a few days later I was at an electrical engineering honor society meeting, and somebody was saying something about taking a survey of electrical engineering students. One of the guys from the study group—and this guy was really bright—turned around and asked what I thought about it. And then I knew, 'This guy respects me. I am one of the people that he feels is on his level.' I try to hold onto that, to know that other people recognize me as their peer.

"At a summer job, I was changing a program that somebody had written. It required a lot of reading. I knew that I was changing a certain parameter that had been considered constant. But I was making the program so it could be changed to be variable for each point to point. It wouldn't necessarily be the same route—it would be a matrix route.

"They had a monthly seminar series. And one of the guys in the group who developed part of this program went up and was talking about how the algorithm worked, about certain assumptions. I could follow it fairly well because my reading made me pretty familiar with the algorithm. So I asked, 'What happens when this assumption fails? How will it change the algorithm?' I didn't necessarily know what it would do, but I knew it was the right question. A few days later, one of the researchers came—kind of breathlessly—into the room and said, 'How did you know? How did you know to ask that question?'

"I figured that I knew what was going on if I was able to ask an intelligent question!"

Michelle
doctoral student, electrical engineering

to believe that negative feedback is justified and to keep quiet. Don't internalize criticism.

• Feed your self-esteem bank. Remember those times when you have asked the "right" question or solved a difficult problem. You'll soon realize you have what it takes to make it. Remember your successes.

• Remember that time will help you feel more confident. As a first- or second-year Ph.D. candidate, it's difficult not to feel hopelessly clueless. As you define your interests and begin work on your own research project, you will undoubtedly develop a stronger sense of yourself and your own abilities. Take one day at a time.

Learning How *Not* to Be a "Nice" Girl

You're certainly at least as smart as the guys in your classes, but the testosterone levels that your male peers boast may be serving to make your playing field a bit bumpy.

The first difference between you and the guys is that you probably argue differently—and that may give them an edge.

A 1990 study showed that women are more concerned than men about the damage that an argument might cause in an interpersonal relationship. In general, men who argue are regarded as "rational," while women who argue are deemed "disagreeable." Therefore, in fields where argument is necessary, women are at a disadvantage when dealing with male peers.

This gender difference (whether true for an individual or perceived as true by others) can also prove to be a disadvantage in working with a male advisor. If a male advisor believes that a woman may react more "emotionally" to criticism, he may not give her the feedback she needs to make her work better. And so the female student is denied an advantage that the male student is not, through no fault of her own (Mapstone, 1990).

Nancy Hutson, who holds a Ph.D. in physiology and is vice president of Strategic Management for the Central Research Division of Pfizer Inc., cautions:

Make no mistake, sex and gender are not the same. Sex is genetically determined, universal, and unchanging. Gender represents those behaviors taught by society and that therefore differ from culture to culture. To a very real extent, our society still mandates that women should be distinct in their behavior from men; more passive, less competitive, and less aggressive. So the first thing we must do is break loose from our gender restraints and, instead, listen to our inner voice. Science, by its very nature, demands confidence, assertiveness, and a competitive streak that will drive one to be the first to make and report an observation.

From "Women Drivers" in (Cecily Cannan Selby, ed.) *Women in Science and Engineering: Choices for Success* (1999).

LEARNING AGGRESSION

"In high school there were predominately boys in the classes I was taking, such as the advanced placement math and science. And what usually happened was that I was very quiet and shy and I got overshadowed a lot. One day in my physics class I realized that I was going to have to start being bossy and stand up for myself. All the guys were crowded around the stations and I couldn't see a thing. I had to climb on top of a table to see what was going on in the experiment. And when class was dismissed and everyone ran to catch the bus, I spent a few minutes trying to work the equipment and figure it out on my own. At that point I just decided I had to be a little bit more aggressive.

"And I think that really helped because when I went to college a lot of times my lab partner would be a guy who would take over. And so I learned to be equally bossy, and my shy personality got replaced by this more bossy one—out of necessity, just to survive. You really can't survive if you're timid."

Lillian
doctoral student, physics

LEARNING THE CULTURE

"I found it useful to see the whole thing as a process of verbal dominance ordering games. Whether you're right or wrong about something usually doesn't matter in a particular conversation nearly as much as whether or not you're able to come out on top in some way at the end of the 'discussion.' It's OK to be wrong, because in the name of science we're all out there trying to find the right answers. But it's never OK to let someone else dominate you in conversation.

"So it's not just the fact of the 'questions' that we face, it's the fundamental style that's most important. It takes practice to be able to go through these 'discussions' and be able to be calm and not back down and not get upset.

"You have to learn how to operate by the rules of a 'foreign' culture. Plenty of women believe in their own intellectual abilities (why would they be there in the first place if they didn't?), but feel like they've been hit by a truck the first time they experience a typical argument in a research group. The aggression levels are high, but they're not meant to hurt. In the minds of the guys it's just part of how the game is played, and part of the reason for being there and enjoying it . . . it's a part of their culture, a part of their pleasure function."

Katrina
recent civil engineering graduate

TIPS FOR NEGOTIATION

Negotiating is a critical skill that you will need from your first day of graduate school on into your career. In her article "Negotiation Advice for Women: How Not to Lose Your Skirt" (Committee on the Status of Women in the Economics Profession Newsletter, Winter 1999), economics professor Linda Babcock presents an interesting argument that women don't "adopt inferior [negotiating] tactics, but don't recognize opportunities for negotiating."

Babcock suggests these tactics for achieving results via negotiation:

1. Recognize opportunities for negotiation. Don't accept everything as the status quo, but identify less-than-optimum situations as opportunities to negotiate. For example, a Ph.D. student wasn't defending until the end of the summer but had wanted to go through the spring graduation ceremony with her classmates. She was disappointed because two male colleagues who also had not yet defended had gotten to go through the ceremony. The problem was that she had not asked, but the male students had.

2. Overcome anxiety over negotiating. Babcock asserts that "women tend to view the conflict inherent in negotiating as jeopardizing" the relationships they have with the person with whom they will negotiate. Men, however, see the relationship and the negotiation as independent of each other. She suggests viewing negotiation as "an accepted and expected activity" that can be done while maintaining one's relationship with the other party.

3. Be sure of what you want to get out of the negotiation. Before you enter into any negotiation, be clear on what goals you are trying to achieve and what is the minimum you will accept. At the same time, be open minded to the other party's objectives and preferences. In all, be assertive, yet creative.

4. Don't view negotiations as win/lose. The "variable-sum" way women typically view negotiations, says Babcock, is a strength that they have over men, who tend to view negotiations as "zero-sum." Women bring a "more cooperative and problem-solving approach and a willingness to understand the interests of the other side to the bargaining table."

Another difference you may have noticed is that guys tend to be "pushier" when it comes to taking over experiments, using the computers, or speaking up in class. But this is not always the case, and even if it is, it is certainly behavior that women can learn and employ as well. The stories on the previous page illustrate how some women see learning to be "bossy" as simply another skill to master along with their studies.

BALANCING COMPETING NEEDS

- When Everybody Wants Your Time
- Take a Break
- Helpful Hints to Balance Your Needs

When Everybody Wants Your Time

Doctoral work is challenging no matter who you are. It pushes you academically, mentally, and socially. Beyond that, it is an intense period of socialization—everybody depends on everybody else. As you learn about your field, you are probably relearning "how to learn." As you learn to "think like a civil engineer," for example, and learn to "think independently," you depend on other people in the field. You depend on faculty members and peers to help you develop an image and sustain a belief in yourself as an academic. Their support and critique of your work helps you to refine your ideas and develop your own line of thinking.

However, *faculty* also need *you*. They depend on you and each other to support, critique, and, in some cases, carry through their research agendas. As you become familiar with their research, you provide insightful direction and critique to their work. As you become more immersed in the field and produce increasingly higher quality work, you become more valuable to both your advisors and peers.

TIP

Don't skip classes. It will take you much longer to catch up than it would have if you had attended in the first place.

Somewhere in there, you still have a private life. That life *also* needs you. You probably once were a social being, before all of this work, and you may even have a boyfriend, husband, or partner. Some of you may even have children. And some of you are responsible for the majority of the housecleaning, cooking, shopping, laundry, and bill-paying. Whether or not a woman is in a relationship and/or has children, she is more likely than a man to feel responsible for maintaining a household in reasonable functioning order. Many men, on the other hand, are less likely to feel such pressure. Balancing these professional demands and personal responsibilities is an area where some women experience more problems than their male colleagues.

Women are also generally socialized to be more concerned with maintaining relationships than men. They may feel a greater need than their male colleagues to establish long-term relationships. The need for a partner may be heightened if they hope to start a family. The need for friendships and support may be no greater for women than for men, but women are more likely to recognize the importance of establishing friendships and feel responsible for sustaining them.

If a woman is partnered and/or has children, she may be more committed to the needs of her family than her male colleagues. Since the majority of women still

---

DISCIPLINE—AND A GOOD ATTITUDE

"When I started back to school, my two children were in elementary school. I still had obligations to my company on small items for my job plus I had to complete my degree program in three years, which had never been done at my university before. I felt a tremendous amount of pressure.

"When you have all of those competing pressures, it takes a lot of focus to clearly understand what you're learning and what you need to learn, trying to do your research on a schedule and to maintain your self-discipline. Even when I didn't feel like doing the research or doing the reading, I tried to discipline myself to find a topic every day that I could work on for three to four hours. But then, sometimes, I would get these energy spurts and could work for 12 to 16 hours straight!

"But I tried to have focus, self-discipline, and consistency in my work habits because I needed to work the hours when the kids were in school because when they came home from school I needed to take them to their gymnastics and listen to them share their stories from the day and their emotions and their feelings. You need focus and self-discipline and consistency and a really good attitude."

Caroline
doctoral student, mathematics

---

carry the primary responsibility for homemaking and childrearing, they may—and often must—invest an equal amount of time outside their profession to meet these personal demands.

Moreover, since research shows that women tend to be less focused on a career as *the* single means of fulfillment, they may feel more ambivalent about the demands that graduate school places on them.

The structure of the academy and its intense and highly competitive environment often make it difficult to balance professional and personal demands. Flexible work schedules and career leaves that make it possible for many women (and men, for that matter) to juggle the demands of work and family are rare in the academic world.

Take a Break

Given the enormous number of competing demands, it is easy to understand why many female graduate students are frustrated. Some students report that their initial reaction was to try and meet all the demands. They wanted to prove that they had the drive to succeed in a highly competitive program. But it is difficult to do it all equally well. If you find that you can't keep up the hectic pace and continue to produce quality work, you are not alone.

Everyone needs time to eat, rest, relax, and be together with friends, loved ones, and family. Each individual's need is completely different. Don't deny your complete self. Of course, you want to work and be a partner and/or friend or mother. If you try to "reprogram" yourself into one line of thinking—and deny one side of you for the other—you'll do more harm than good. Instead, try to develop strategies to deal with your conflicting time demands.

One strategy involves setting priorities and saying no. Your decisions on what to prioritize are important. Know what your department wants ahead of time so that you can fulfill those demands more efficiently. At the same time, you can attend to some of your important personal needs.

Setting priorities also involves keeping personal demands in perspective. Sometimes we use our personal life as an excuse for not doing what is actually important. As you adjust to the demands of your hectic life and become more comfortable in the work environment, you'll learn to set priorities that help you maintain a

---

LEARNING TO FOCUS

"For a long time I was just struggling—and getting nowhere. I spent too much time networking and being with people, because I could do it easily and it was fun and engaging. But I spent too little time on my academic work.

"I started losing confidence in myself. I had thought graduate school would be about finding new ideas and having somebody to talk to who was excited about these ideas. The reality was more like, 'Buckle down and generate your own excitement.'

"Nobody paid any attention to me. At first, I thought that was great. I could take a day off here and there and nobody was hanging over my shoulder. Then I started thinking, 'You have to finish sometime. And you're not getting anything done. You don't even know where to start looking!'

"When I realized that my time was not going into getting my own work done, I had to extricate myself from my cobweb of commitments and expectations and learn to say 'NO.'

"It was a painful operation—and I can't say exactly how I did it, but step by step I distanced myself. And I told my advisor that we had to keep in touch a lot more than we had in the past.

"Then I just spent a lot of time struggling. But once I found out how to focus on my work, I became motivated and established a schedule.

"And finally, I had the feeling of breaking through the wall and seeing the light. It seemed so obvious. I have been told that this is normal, that people go through this all the time, that they bang their head against a problem and can't seem to see how to get out of it, and if they just stick with it, something happens. The problem is posed in a different way, or rewritten, or they imagine it differently, and suddenly things seem easy."

Lori
doctoral student, physics

balance. In time, you'll learn to both separate and integrate your personal and professional lives. One or the other will take priority during deadlines or emergencies, but the two can and *should* be balanced whenever possible.

Helpful Hints to Balance Your Needs

- **Schedule your free time like you do your work time.** Make dates with yourself or others to exercise, and have dinner with friends or family. If you don't write it down among all of your other commitments, fun and relaxation might get pushed aside by school work deadlines.

- **Personalize your work style to go with your lifestyle.** No one can come up with a perfect prescription for balancing. And remember, making choices requires constant adjustment to unforeseen conditions.

- **Have a plan but celebrate the short term.** Understand and articulate to yourself as clearly as possible your goal(s) and your reason(s) for getting an advanced degree. Then, as you work out short-term goals over the course of your graduate career, keep in focus what you want to achieve.

- **Maintain realistic expectations about what you can accomplish during a particular time period.** Set your short-term goals and plan your daily and weekly routines carefully. Too many graduate students set themselves up for failure by assuming that they can manage a Herculean schedule long term. Stick with what is reasonable rather than what is enviable.

- **Set up your timetable as much as possible so that it functions according to your time clock.** We all have peak hours for working, so, for example, if you work best during the morning, set aside a block of study time during this period of the day rather than at another time. Use your biological rhythms to determine what activities to schedule at what times. Capitalize on your own prime time.

- **It is equally important to balance your academic/intellectual efforts with the relationships that provide meaning and depth to your life.** Your life will be infinitely more rewarding if you can plan at least a little time to be spontaneous or create some space to react in the face of emergencies and deadlines. Tune into your "whole" self.

- **Understand that you can have a lot of what you want, but you can't always have it at the same time.** There will be times when you need to put some things aside while attending to others. The trick is learning to put some things aside just long enough to accomplish the immediate task at hand, but not so long that you begin to feel severely deprived.

- **Set limits with both your personal and professional time.** When you need to get work done, limit your interruptions. Focus on the job at hand. At the same time, when you're with your family or on vacation, don't talk about work!

Deidre Meldrum

BALANCING WORK AND PLEASURE

"While at Stanford I got married and was involved in a lot of activities in addition to working on my Ph.D. My husband was also a Ph.D. student. I think this helped us both a lot because we understood when each of us had to spend many hours, day and night, working on our research.

"Nonetheless, we balanced work and pleasure. I suppose I took longer than necessary to complete the Ph.D., but in hindsight, I'm very glad that we did it this way. Being in graduate school was one of the most enjoyable times of my life not only because of what I learned at school but because of the friends I made and good times I had with them. As a graduate student we went on a bicycle ride for two hours every day with friends and went on long rides on the weekend with friends and/or my husband's faculty advisor. We went on week-long trips cross-country skiing in the Sierra Nevada Mountains. I also worked out regularly swimming, doing aerobics, and rowed on a crew team. What was important in being able to do these extra activities was that I had certain times set aside for research. Overall, I would say that we played very hard and worked very hard.

"Looking back on these experiences, although graduate school is demanding and requires a great deal of commitment and work, one is fortunate to be able to concentrate on courses and your own research. Now that I'm a professor I have courses, research and a zillion other demands on my time (students, committees, family, etc.). I have to be more efficient and organized than ever."

Deirdre Meldrum
Associate Professor of Electrical Engineering,
University of Washington, Seattle
(Ph.D., Electrical Engineering, Stanford University)

- **No matter how hard you try, things may just be unbalanced.** Your needs may become so buried in all the "to do" lists that you have no idea what they are! Don't worry, everyone experiences periods of productivity and chaos. If you're feeling overwhelmed, talk it out with friends and colleagues. Just be careful whom you choose because women are often viewed as "stressed out." Find a group of confidants whom you can trust. They can help you reset your priorities and set things straight.

- **Keep your professional life in perspective.** "Proving" yourself does not necessarily mean pleasing everybody. Do the best you can, be supportive, helpful, and hard-working. Beyond that, no one can expect anything else from you.

4

THE FINAL DESTINATION: AFTER THE PH.D.

"Keep in mind that mental toughness, or perseverance in the face of obstacles, is fundamental. In fact, I could say that I got my start in science out of sheer stubbornness. When I went to high school, girls simply were not allowed to take physics. What's more, my high school chemistry teacher told me I'd never make it in chemistry—because women couldn't. That angered me but also galvanized me. I had begun to see science as a way to understand the world and a way to make my way in the world. . . .

"The tough challenges facing us as women in our treks through science and education are far from over. But as scientists and engineers, we know the value of both reason and imagination. And . . , we know perseverance can carry us on great journeys."

Dr. Rita Colwell
Director, National Science Foundation
(Ph.D., Marine Microbiology, University of Washington)
From her speech, "Trailblazing: One Woman's Trek in Science,"
at the Association for Women in Science—Alaska Chapter Luncheon,
October 26, 1998

THE JOB SEARCH

- Guidelines for Finding a Job
- Applying for the Job(s) You Want
- Interviewing
- What You Should Ask
- Negotiating the Job Offer

Although you may still be in graduate school—and getting through each day is the first thing on your mind—it's important to keep an eye on the future. It's never too early to start preparing for your first job.

Finding a job depends on any number of variables: the time of year, your specialized field, where you want to go, whom you want to work for, and, most importantly, who is hiring. There are other variables to consider, including looking for a job in the private sector or academia, nonprofit or public sector, delaying the job search to have a family or get married or considering further study as a postdoctoral candidate. Whatever the case, looking for a job requires serious thought and a great deal of work. One thing you can affect is your level of preparation. Finding and successfully landing the job you want depends entirely on how much effort you put into a job search.

It can take some time to find a job. In physics, for example, it can take six to nine months, depending on the field. And it can be very exhausting to job hunt while finishing your thesis. You can, however, have some of the job-hunting preparations out of the way before you are ready to completely devote yourself to the job search. For instance, you can have your curriculum vitae ready to send out to potential employers, and you can start a list of contacts.

TIPS FOR INTERNATIONAL STUDENTS

Timing is critical for international students as they finish their theses and begin looking for a job. One foreign student advisor suggests that if you are an international student who wishes to work in the United States you should:

- **Start working on the job search at least a year before you expect to graduate. Talk to your university's foreign student advisor about filing for a work permission permit, which can take three months or more to be approved by the Immigration and Naturalization Service.**
- **Complete your thesis and defend early in your last semester; then work on your job search the rest of the semester, leaving time at the end to put the finishing touches on your thesis. (Be sure to discuss this plan with your advisor, however, since he or she may prefer that you complete your thesis as soon as possible.)**

Guidelines for Finding a Job

While there is no sure way to guarantee a perfect match, a good plan can help ensure success. Simply put:

- Apply for the jobs you want.
- Interview well.
- Negotiate a satisfying job offer.

PREPARATION IS ESSENTIAL

"In my next to my last year at Stanford, I started investigating opportunities for a career in academia. I researched a wide variety of universities and obtained as much information as I could about their courses, faculty, programs, etc. I prepared a c.v. and tailored each one to the university/department I was applying for. I applied to a lot of universities in early fall of my last year and was invited to interview at seven universities. The most important part of this process was being very well prepared for each visit. This included having a very well-polished research presentation, knowing about the research of individual faculty where I was interviewing, what courses I could teach, what I could contribute to their department, and being able to explain very clearly my own research and interests. I was fortunate to have offers at five universities, which put me in a good negotiating position. It was actually a tough decision to choose where to go but I weighed factors such as opportunities, environment, student caliber, colleagues, resources, needs of my family, etc.

"I am grateful that I worked hard on the negotiating phase of the package because that is what provided my initial research funding, salary base for the rest of my career at my current university, research students, and the time that helped me to get my research program going. I'm also glad that while I was in graduate school I networked at conferences, presented my work, and made sure that others knew what I was doing and was interested in."

Deirdre Meldrum
Associate Professor of Electrical Engineering,
University of Washington, Seattle
(Ph.D., Electrical Engineering, Stanford University)

This simple list does not do justice to the complexity of each of these tasks. And, of course, no list is set in stone. Depending on what kind of job you want, whether it's in industry, public service, academia, other nonprofit areas or consulting, and what kind of person you are, the list—and the work involved—can vary tremendously. In the rest of this chapter we outline the steps and provide a checklist for navigating a successful search.

Applying for the Job(s) You Want

Assessing Your Skills. You can't apply for a job until you know what kind of job you want. Although this statement sounds simple, figuring out what you want can be an enormous challenge. Uncovering available jobs is part of the task; finding a job that will satisfy you both personally and professionally—in short, a job that will make you happy—is the other part.

Take inventory of your skills and what you have been doing in your graduate school career. If it requires writing down on a sheet of paper what you can do and what you like to do, then by all means—do it! Unless you're considering a 180° turn, you should be reasonably certain of the direction you want your career to take. Many of you will pursue careers in academia, becoming research fellows or professors—more than likely in the field you've been studying for your entire collegiate career. Others will try for jobs in industry as scientists, engineers, or research and development professionals. Some may begin or join start-up companies. Many will pursue all of these. Whatever the case, you should know what your capabilities and special skills are and what you like to do before launching your search.

Discovering What's out There.   The next step in finding a job is determining what jobs are available. This requires understanding the market you want to enter and how it functions. In academia, the most common ways to learn about job possibilities are to ask your advisor for contacts, read professional journals in your field, and attend national conferences. Available jobs may also be listed in publications like the *Chronicle of Higher Education,* in the human resources departments of most colleges and universities, and on the Internet. Your university career center may also have listings.

For jobs in industry, nonprofit organizations, or the government, recruitment methods can vary. Some major national companies actively recruit students, including master's and doctoral candidates, on campus. Some career centers also have files on most major companies, where you can research prospective employers on your own. Jobs in business or industry are listed in some national publications and can also be found in the classified ads in major newspapers. These methods all take a great deal of time and energy.

Networking.   A crucial means of discovering employment opportunities is using your connections in the field or "networking." If you are looking for a job in higher education, your advisor will probably know of opportunities in your area of study and may have names of people to contact. If she or he does not make suggestions, it's important to at least ask. Although you may feel like you're using someone for your own personal gain, asking about employment opportunities is as common as asking about the weather. Most people are willing to help.

If you are in the uncomfortable situation where your advisor does not offer to help you with your job search or where you have decided to pursue an alternative career path that she or he does not support, be prepared to seek help from other sources. In some cases, you may be able to convince your advisor of the merits of your choice, and everything will be fine. In other cases, you will have to find other persons who can suggest career opportunities and speak effectively about your work to prospective employers.

Remember that your network is much more than just your advisor. Talk to your committee, other faculty, or other professionals in your department and field. Perhaps one of them attended the school you're interested in or knows something about a certain company. Speak with other graduate students, family members, friends, or even friends of friends. No one should be left out of your circle; you never know who can give you a promising lead on the perfect job.

Do Your Homework. Whatever your interests, once you have heard about job opportunities from your various sources and contacts, make sure to do your research and find out everything you can about the position. Write letters, ask questions, make phone calls, request materials—whatever it takes to find out as much as possible. Ask questions about the nature of the job, what the department or division is like, how new people are treated, what kinds of opportunities are available. Don't be bashful! After all, you want to have the necessary preparation to make an informed choice.

A Curriculum Vitae. You probably already have copies of a resume lying around, but make sure that you invest the proper amount of time into perfecting your resume for the application process.

Most people in academia compile a curriculum vitae, or c.v. The c.v. details your academic history, complete with research focus, dissertation topic(s), history of published articles, professional experience and presentations. If you're unsure how to go about preparing a c.v., ask your advisor or other people in your department for samples and help. Your career center may also be able to provide background literature and help.

If you are preparing a traditional resume, you may also want to consult the same sources as above. There are hundreds of resume preparation books, complete with samples, that you can use to help you through this process. Resumes take various forms and there are many debates as to what should or should not be included. Once you have decided on a form and have ironed out the content, have it checked by various friends and professionals for accuracy and effectiveness.

Whatever you use, make sure that your c.v. is free from errors and that it accurately depicts your strengths, experience, and accomplishments. Never "pad" your resume with fluff, but don't be modest either. Never falsify your c.v.: most employers do check references at some point in the hiring process, and if you are caught in a lie, even a little white one, it could doom your prospects for future employment.

Applications. As soon as you have a solid job lead, decide if you want to pursue it. By doing your homework in the information-gathering stage, you should be able to eliminate those opportunities that aren't right for you. Don't waste your time and effort applying for a job that you wouldn't take anyway; on the other hand, you don't have to be absolutely sure about a job to apply for it. If your qualifications and experience do not exactly match a position description, or if you're unsure about a particular location but it seems possible and interesting, you should try applying. Give yourself every opportunity. You can always make a decision later or be pleasantly surprised by an invitation to interview for a position! Too many people eliminate themselves if the "match" isn't exact. But many positions are flexible; so if you think you're interested and can offer something to the employer, give it your best!

How you apply for a job is important. Use the resources available to you, the job description, published information, and what you've discussed with people close to the job to write your application letters. A carefully crafted letter can sell your application.

Interviewing

While the interview process depends on what kind of job you're interested in and where you apply, following a few general principles can be a real help.

Always come to the interview prepared. Know as much as you can about the position, the organization, and any other pertinent information. Be ready to talk about yourself and your strengths as they relate to the specific position and any other topics about which the interviewer may ask you. Practice thinking and speaking on your feet. Interviews can take many forms, and you must remember you are "on" at all times.

Practice all of your social graces and take care with your appearance. Although the job may require casual attire, most people recommend that you interview in a suit. Take the time to show a potential employer that you care about the position enough to look and act professionally. Even your body language can demonstrate how you feel about the situation. Obviously, don't smoke, chew gum or pick your fingernails. Keep in mind that an interviewer is often looking for a reason not to hire you.

The entire interview process is designed to showcase your qualifications for the position. It also provides a time for you to gather information. You want to make sure the job is right for you, that you "fit." Try not to be intimidated, afraid, or nervous—although a little anxiety is normal and expected. If you approach an interview with the right attitude, your experience will be positive.

Academic Interviews.   In academia, you may first interview at a conference and then be selected for an on-site interview. If so, the initial interview is a short screening interview to determine if you have the surface skills and personality for the job. The interviewers have most likely seen your c.v.; perhaps they have talked to people who know you and your work. In this kind of situation, be prepared to summarize your work and skills in what may be a very short period of time. Try to make the best aspects of your personality and work habits come across through the discussion, and quickly and effectively discuss your plans and goals.

A major challenge in this kind of interview is not having enough time to adequately discuss yourself or your work. Your personality may take a back seat to just getting through research accomplishments, and the interviewers may not have a chance to understand who you are as a whole person. Try to answer questions as thoroughly and consistently as possible. Although you may feel that you're leaving a lot out of the picture, if you successfully get through this stage, you'll have more time to discuss yourself on site.

Academic interviews typically vary with the goals and priorities of the department to which you're applying. Doing your homework is definitely necessary. If you know what you're getting into, you'll know how to prepare for the interview and what kind of information to bring. However, as a rule, most schools will focus on your research, your teaching experience (especially if English is not your first language), and the question of whether teaching is important to you.

Typical interviewing situations include interviewing over the phone or via computer, interviews with more than one person, or interviews in which you're asked to make a presentation or prove your teaching or research skills. Again, good preparation is critical. Be prepared to prove your qualifications for the job in any kind of situation. It's all right to ask before the interview occurs if you should bring any kind of materials, writing samples, or copies of published articles. You can even ask if you'll be tested or expected to perform. It can be a real help to talk to other people about their experiences. Don't be shy.

The Job Talk. At most academic interviews, you will be asked to give a job talk in which you present a scholarly overview of your research. One engineering professor who has both given and attended many job talks offers these suggestions:

- **Know your audience.** If the audience is a mixture of those who are well acquainted with your specialty and those who are not, you must make sure that everyone in the room is able to understand at least some part of your talk. If you don't use the proper jargon, the specialists will think you're not one of them; but if you speak only in jargon, the nonspecialists won't know what you're talking about.

- **Show the big picture.** You will potentially be working with both specialists and nonspecialists, so you want to show your technical depth while indicating how you will fit into the department.

Business or Industrial Interviews. Most business interviews are very similar to academic interviews, and the same rules and suggestions apply. Potential employers are trying to determine if you are the right person for the job and organization, since your specific assignments may change over time. You may be asked to prove yourself in any number of ways, from answering unusual questions to taking a skill assessment test.

Behavioral interviews are all the rage these days. In these kinds of interviews, you will be asked how you handled certain situations in past. This interview approach is based on the idea that past behavior may predict future performance.

Whatever the interview style, always do your homework about the organization and the job and be prepared for anything they might ask you.

Unusual Questions. Some potential employers ask unusual, or even offensive, questions. Sometimes they are trying to determine an applicant's level of commitment to the job. Women, in particular, may run into this type of questioning. In some situations, questions may be difficult to answer and may even be illegal to ask, for example:

- Are you married or planning to be married?
- Do you have a family?
- Are you pregnant or planning to become pregnant in the near future?

- Do you have any unusual conditions that would prohibit you from doing this job effectively?

- Do you smoke?

(Under Title VII of the Civil Rights Act of 1964, it is even illegal for an interviewer to ask about your ethnic or national origin, such as where you were born or even if you are a U.S. citizen—*unless they ask all persons who are being interviewed for the job.* This stipulation, however, puts the burden of proof on the person suspecting discrimination and may be difficult to pursue. To make matters more difficult, it is not illegal for an employer to discriminate on the basis of a potential employee's noncitizenship once the employer is aware of the person's status.)

Knowing how to answer these kinds of questions is not easy, and even experts disagree on the best approach. Some suggest not answering such questions directly, but instead answering the question behind the question: "I recognize the time required to be successful in my field and am committed to doing excellent work." Others suggest asking why the question is relevant to the job and basing the answer on the interviewer's reply. Whatever the case, use your best judgment in the situation, and carefully gauge your own feelings about answering the question. If you feel you've been asked something completely out of line, you could consult people you trust for advice and might even consider speaking with the human resources department of the organization. Whatever you decide, think hard before you react.

What You Should Ask

As a general rule, it isn't acceptable to discuss salary or benefits on the first interview. In an academic search, you can bring up issues about grant funding, facilities, or graduate student support, but directly asking about salary or benefits is considered inappropriate. The same is true for many organizations. Ideally, what motivates you should be consistent with the needs and goals of the organization.

Ask about the duties of the job (if they aren't adequately explained), the climate of the school/ department or company, outside duties, and any expectations.

Many people may also want to inquire about the possibilities of a dual career move. Some universities or companies will try to find employment for your spouse or partner if you accept a position. Most people, however, feel it is advisable to raise the two-body problem only after it is clear that the job offer is pending.

Try to gauge the way the interview is going before you ask questions. If you feel comfortable with the interviewer, your questions may take on a more personal tone. If not, reserve your questions for other people who may know about the company or other interviewers you may encounter throughout the process. Don't hesitate to ask an interviewer for suggestions of other people with whom you can talk about the job or company or school. Always have at least a few questions for the interviewer. It shows your interest in the job and in the process.

Negotiating the Job Offer

If you've gone through a successful application and interview, it's only logical that you'll receive a job offer—and hopefully more than one! After the initial thrill wears off, there are several factors to consider in accepting or rejecting an offer (such as salary, benefits, and moving expenses). Most likely, you'll be involved in negotiation with your prospective employer.

Most offers are extended over the phone or through a letter. Ideally, your prospective employer will use both methods. To be on the safe side, do not accept any offer until you have seen it in writing. If your employer doesn't mention sending a written copy of the offer simply request one. Alternatively, you can write a letter to the employer listing what you understand the employment package to include and state, "if this is not correct, please clarify." To be safe, you could have this letter notarized.

In all negotiations with your employer, remain composed and relaxed, yet firm. Remember, these are potential colleagues, and you want to operate in good faith. Assume that the party extending the offer to you is also operating in good faith. She or he is eager to fill a position and has decided that you are the best person for the job.

TIP

Exploding offers (offered early in the interviewing season with extremely short decision deadlines) can be dangerous. Deal with these by scheduling these early interviews as late as possible and ask to have the response deadline extended. This will open up your opportunities with other employers. Let other, nonexploding-deadline employers know that you are under pressure.

(From "Negotiating an Offer," a Graduate Student Seminar by Laurie Weingart, associate professor of organizational behavior at Carnegie Mellon University)

After an initial offer has been made, make sure you understand everything by asking questions. It's perfectly all right to ask for more time to "think it over." In fact, you should never accept an offer outright before you've had time to fully consider it.

If all the terms are perfect, congratulations! Call your employer and accept the position. Write a thank-you letter for the offer, and include your starting date. Your job search is over.

If you're not so lucky and you have questions or simply would like to negotiate on certain terms of the offer, map out the issues you want to discuss in advance. For example, with academic jobs, you may want to discuss summer salary, a start-up budget for research, computer or other equipment, or teaching load. It's very important to identify the issues you want to discuss with your potential employer before you actually start negotiations. Have a game plan and be willing to "win some and lose some." Decide which issues are truly important to you—what your priorities are and how the issues rank in order of importance. Then consider what your employer thinks is most important. By articulating your own needs and anticipating theirs, you can identify areas for give and take. Some items of the offer are more than likely not negotiable, such as benefits packages. Most companies and universities have standard packages. Other factors, like moving expenses, bonuses, starting date, and salary may be flexible. Speak with faculty members, other employees, or human resources personnel to find out what the latitude is in the organization regarding these components of a job offer. If you're going to ask for more money, it can help to have solid information to back your request. If you're an academic candidate, check and see if the school is unionized. Most unions set salary scales.

Historically, many women have been reticent to negotiate parts of a job offer, especially their compensation package. Even if you feel that money is not your priority, if you have a green light to negotiate, go for it! This may be one of the most powerful positions of your career. The organization is interested in hiring you, and you may just get all or most of what you're asking for. Stop to think about it: what have you got to lose? Since your starting salary is the basis for the rest of your raises throughout your employment, it is to your advantage to secure the best offer you can. Just don't identify issues as nonnegotiable if they are. It is foolish to say you will not accept the job offer unless "X is offered," if you would take the offer regardless.

If you're considering other job offers, it doesn't hurt to mention them, but be careful in divulging details. If one position has all the right features except one and another place is very strong in that missing area, it may be wise to say, "Organization X is offering me four weeks of vacation as opposed to your three. I'd much rather be part of your team . . . is there any way of working this out?" You just might get what you're asking for.

Making It Final.   For job offers in academia, other nonprofits, public service, or industry, make sure you know your time limit for replying. In some cases, several weeks may be acceptable; in others, the standard is a week or even less. Don't let a prospective employer rush you, but know the rules. If they want you, they are usually prepared to wait at least a little while.

Once you have worked out all the details in the offer and you have an acceptable agreement—whether it's an official letter or contract—go ahead and "sign on the dotted line." If you've received a written draft and still have questions or notice missing details, do not hesitate to call your employer with questions. Request a new written letter of agreement.

To be safe, do not reject a job offer until you have accepted another one. If you are sure you do not want to accept, let the party know as soon as possible. When rejecting an offer, call the person with whom you've been dealing personally. Explain your decision concisely and politely and then follow up with a letter repeating the same information. You never know whom you may run into down the road, so don't "burn any bridges."

Potential Questions for Late Interview and for Offer Negotiations

BENEFITS

- Is my health insurance a set package, or can I choose options? What is covered? Who is covered?
- Do you offer reimbursement for continuing education or training?
- Can my family have reduced or free tuition?
- Do you offer subsidized loans for housing?
- Can you help me buy or sell a house?
- Is company or university housing available?
- Do you help with spousal placement (relocation)?
- Will you pay for a trip to house-hunt and take care of other necessities?
- Do you have day-care facilities?
- What are your policies for family leave or sabbaticals?
- What is your vacation and holiday policy?
- Do you pay moving expenses? If so, what is the flat amount? What is the percentage?
- What are the arrangements for parking?

SALARY

- What is the starting salary offer?
- Is salary negotiable?
- Are you offering a 9- or 12-month contract?
- Is there an annual year or fiscal year salary?

For Academics

- Is there summer support? How much?
- How long is summer support guaranteed?
- Do I teach in the summer or receive a research stipend?
- Are the three months of summer salary tied to research productivity?
- Can I get summer support the summer before I begin teaching?

WORK LOAD

- How much committee work is involved?
- What is the usual teaching load?

- Are the courses graduate or undergraduate level?
- Will my classes have teaching assistants?
- How many new courses will I teach or have to prepare for each year?

PROMOTION/TENURE

- How long before promotion/tenure review?
- Could I start on the tenure-track before finishing my dissertation?
- Would I continue on the tenure-track if I take family leave?
- Who is involved in the tenure review process?
- How many publications are important?
- Are conference papers considered?
- How are teaching scores evaluated for tenure?

BONUSES

- Is there a start-up bonus, and am I eligible?
- How much money is available for start-up?
- What can I use the money for?

PERSONAL/OFFICE NEEDS

- Will I have my own computer equipment, and what kind?
- What kind of software/hardware support is offered?
- What software is available on site?
- Will you provide additional computer equipment/software if necessary?

LAB EQUIPMENT AND FUNDING

- How is lab funding allocated?
- Will my projects be under senior faculty, or will I have my own funding?
- What is the expected timetable for starting research projects?
- What kind of facilities will I have access to?
- What kind of equipment can I get?
- Can I get the necessary support for my research projects?
- What kind of support is there for research assistants?
- Are assistants provided, or will it be my responsibility to find them?

DON'T WORRY!

"It takes a lot of introspection and sometimes it takes a long time, but no matter what your family or personal situation is, if you did the right thing in graduate school, if you pursued something that was important to you and you did a good job of getting trained, then chances are you'll be able to do something that you really want to do afterwards."

Lin
recent robotics graduate

APPOINTMENT

- What is my contract review cycle?
- What are the conditions of review?
- What standard of work ensures contract renewal?

MAYBE A POST-DOC FIRST?

Forty-three percent of students graduating with an engineering or science doctorate in 1995 planned to pursue post-doctoral study.

Why are post-docs so popular? A post-doc is a good opportunity to establish your identity in your field and learn to write proposals. Especially if you are planning to enter academia—where you will have teaching and committee responsibilities in addition to your creative work—doing a post-doc is a good place in between where you can really concentrate on your research. In some fields, such as the lab sciences, completing a post-doc is generally expected as a qualification for a tenure-track faculty position.

A biology professor who supervises post-docs comments as follows: "One does a post-doc to establish a reputation in a growing, emerging field and to become familiar with the ideas, methods and people in that field." Furthermore, he adds that the post-doc experience "complements the graduate training where one's subject of research is less important than simply learning how to do 'good science'. . . . [L]earning how to do an experiment is often less important or difficult than learning when to do it."

A post-doctoral position can be a useful stepping stone from graduate school to a permanent job, but "beware getting a post-doc when you could be getting a faculty position," advises a professor who once held a post-doc position in civil engineering. "At some places, a post-doc is cheap labor, and the department is getting a quality young Ph.D. to do real work for practically nothing. Plus, this is one more year where you aren't in a 'real' job—namely, a real academic faculty job. Think harder about your *career,* and don't just look around for a job."

In the "old days," your advisor might help you find a post-doc at another institution. Today, however, given some students' family obligations and the fact that departments have invested a lot in the student, many stay on at their graduate school. Seeking a post-doc position outside of your current department is the best route; otherwise, you may tend to keep your student status in the advisor's mind and may not have the opportunity to establish your own identity. Change for change's sake isn't essential, but even if you are happy with your current environment and your current research specialization, it's generally worth at least a look elsewhere.

Seeking a post-doc position is done pretty much the same way as seeking a regular job. Talk to your advisor and other department contacts, network at conferences, check ads in journals and the *Chronicle of Higher Education.* You should also consult the offices and publications of specialized organizations in your field. Some professional organizations (such as the American Physical Society) will even distribute your resume for you.

Identify (by reading, networking, and attending seminars and conferences) those in emerging, interesting subfields who are currently getting the most funding and then contact them to request an interview.

You should follow up on ads for post-doc positions with a telephone call or e-mail message to clarify certain issues about the position. Short advertisements often fail to detail the responsibilities of fellows, such as teaching expectations, grant-writing, and participation in existing research projects.

You may be surprised that grades don't count much when applying for post-docs. Most impressive to potential employers are letters of recommendation and research accomplishments—especially publication in prestigious journals.

At the interview, you should not only impress the interviewer but also use the opportunity to check out the lab, department, city, and whether the whole environment will be a nurturing one for you. If you are looking for a mentor, talk with graduate students and other post-docs about how available faculty are for that kind of relationship.

One year is a good tenure for a post-doc when you are continuing at your graduate school; two years is optimal, three at the most, for other cases. A couple of years allows you to get comfortable in your position and get some papers out.

But personal circumstances and the job market may impact this ideal length of tenure. A major growing problem is the inability to find a permanent position after the post-doc. Thus, more people are staying in post-doc positions longer than they should. On the other hand, if you are in a post-doc position that fits your lifestyle (such as when you are starting a family) and your professional needs, it's a unique opportunity that you should take advantage of.

"Being a post-doc is great," says a research scientist in the physics department at a major research university. "For someone who loves research, it is the best job in the world. You should view every post-doc opportunity you are offered as a unique gift, however, since it is highly unlikely that any post-doc nowadays will lead to a faculty job. Most of my graduate school and post-doc friends have left academia (typically after their second post-doc) and are now really enjoying the challenges (and pay checks) that the 'real world' has to offer. So start daydreaming about alternative careers as early as possible—something that also acts as useful therapy on those dark days when nothing is going right with your research."

WHAT IT'S LIKE OUT THERE

- The General Workplace
- In Academia
- In Industry
- In the Public Sector

The General Workplace

After an average of 10 years of college, you are finally ready to go to work.

You have a lot going for you. With a doctorate in science, you are among only 15% of women thus educated in the total science labor force; with a doctorate in engineering, you are among the even more elite, since only about 5% of both women and men in that field have doctorates (National Science Foundation, 2000). (See Table 4.1.) You will most likely start out at a very decent salary, and you can look forward to earning a comfortable living throughout your career. (In 1995, the NSF reported a minuscule 1.5% unemployment rate for both men and women with doctoral degrees in science and engineering [National Science Foundation, 1999].) You will enjoy the relative autonomy that comes with having earned the highest degree you can in your field, knowing that you are truly a professional in your area. You will also have the flexibility to move among different sectors: you can work in academia, industry, the public sector; given your area of expertise, you could even do some consulting.

There are still some discrepancies in numbers of women compared to numbers of men in the engineering and science workforce. As in graduate school, you will still be very much in the minority among your male peers, whether you are in academia, industry, or the public sector. While women in 1995 made up 46% of the total U.S. workforce, they accounted for just 22% of the science and engineering labor force. And dollar differences might be the case, too: An NSF report shows that women in science and engineering with the same level of education earn less than men within their age group, though the report does point out that women doctorates are generally younger than their male counterparts and so have less experience and command lower salaries (National Science Foundation, 2000).

But numbers—good or bad—don't tell the whole story. There are still some things about being a woman in the science and engineering workforce that can't be quantified. There is no easy way to measure, for example, how well you will get

Table 4.1 Scientists and Engineers in the Labor Force, by Occupation, Sex, and Highest Degree: 1995

Occupation	Total		Bachelor's		Master's		Doctorate	
	Women	Men	Women	Men	Women	Men	Women	Men
Total science and engineering	728,000	#####	385,300	#####	237,600	678,100	95,800	329,900
Computer/mathematical sciences	279,100	687,100	189,600	445,800	79,500	194,100	9,800	44,800
Computer science	240,900	613,900	178,500	426,700	58,500	166,000	3,600	18,900
Mathematical science	12,700	25,800	5,400	11,200	5,700	8,600	1,600	6,100
Postsecondary computer/ mathematics teachers	25,500	47,400	5,600	7,900	15,300	19,500	4,600	19,900
Life sciences	108,800	202,700	47,800	76,100	28,600	36,600	27,700	76,700
Agricultural/food science	11,600	32,600	7,600	17,300	2,500	7,100	1,500	8,200
Biological sciences	69,900	102,900	32,500	38,800	16,100	17,500	17,800	41,700
Environmental science	2,900	17,800	2,000	12,000	800	5,100	100	700
Postsecondary life science teachers	24,400	49,400	5,800	8,000	9,300	7,000	8,300	26,200
Physical sciences	60,900	220,900	34,300	96,700	16,900	52,900	9,800	70,900
Chemistry	29,700	84,200	19,600	46,700	6,100	14,800	4,000	22,600
Earth/geology/oceanography	13,300	59,800	6,700	30,400	5,300	20,100	1,300	9,300
Physics and astronomy	3,500	26,600	1,000	6,400	1,600	6,500	800	13,700
Other physical sciences	4,700	12,600	3,100	6,200	1,300	3,900	300	2,400
Postsecondary physical science teachers	9,700	37,700	3,800	7,200	2,600	7,600	3,300	22,800
Social sciences	160,600	160,900	32,000	29,600	81,300	56,700	43,000	70,900
Economics	9,600	24,500	4,100	7,200	4,000	11,600	1,500	5,600

Political science	4,000	5,000	2,100	3,000	1,600	1,100	300	900
Psychology	102,700	66,300	16,800	9,100	60,300	29,600	22,800	25,600
Sociology/anthropology	7,700	8,600	3,300	4,500	3,200	2,300	1,300	1,800
Other social science	6,300	6,400	1,800	2,000	2,200	2,700	1,400	1,600
Postsecondary social science teachers	30,200	50,100	3,800	3,800	10,100	9,400	15,700	35,300
Engineering	118,600	######	81,700	849,800	31,400	337,800	5,500	66,600
Aerospace engineering	4,600	70,600	2,900	41,000	1,500	25,300	200	4,200
Chemical engineering	9,400	63,500	6,300	40,300	2,200	17,700	900	5,500
Civil engineering	18,500	184,500	14,000	131,700	4,100	48,600	300	3,600
Electrical engineering	23,300	346,300	16,400	233,400	6,300	98,400	500	13,400
Industrial engineering	9,600	62,300	7,100	46,800	2,400	14,700	100	800
Mechanical engineering	15,000	246,200	10,300	185,100	4,000	53,300	700	7,400
Other engineering	35,900	254,000	24,500	166,300	9,800	72,200	1,600	15,200
Postsecondary engineering teachers	2,400	29,100	200	5,100	1,000	7,500	1,200	16,400

NOTES: Because of rounding, details may not add to totals. Total includes "professional and other degrees."

Scientists and engineers are defined in terms of field of employment not degree field.

Source: National Science Foundation. Women, Minorities, and Persons with Disabilities in Science and Engineering: 1998. Arlington, VA, 1999 (NSF 99-338).

along with your male colleagues, and they with you, or what walls or ceilings you may encounter in your day-to-day work. In her book *Has Feminism Changed Science?*, Londa Schiebinger reports that "Barriers are maintained in part because even well-meaning men and women tend to know more persons of their own sex and to think of them first when putting together committees, conferences, or other work groups" (Schiebinger, 1999).

Karin Rabe, currently the only woman in the Department of Applied Physics at Yale, comments that being one of a few women (or the only woman) in a group puts pressure on you to "prove that you deserve to be there" and find out the information that gets filtered through the male grapevine. You also have to figure out how to "interact with your male colleagues of various generations in a productive and professional way" (Selby, 1999). All these lessons certainly weren't taught in any of your classes.

Any job is going to have its ups and downs, but you might just have to face a few more obstacles in your chosen field. You may find yourself working with men who perhaps haven't had female peers before, and you may face stereotyping by faculty and/or students if you work in academia. You will most likely be asked to serve on committees that desire a woman's perspective, but be overburdened with committee responsibilities because you are the only woman in your department. You may even face sexual harassment, either personal or environmental.

You may or may not encounter any of these obstacles, or you may have already encountered them in your education and easily dealt with them. But they are some of the realities of working in what many still feel is a male domain.

Visibility vs. Invisibility.   Though at times you feel very visible (such as when you're speaking to a room full of your male colleagues) or very invisible (as when you're *trying* to speak in a room full of your male colleagues), there are pros and cons to each of them. You want to blend in—be invisible —so that you can get along with your co-workers and get your work done. But you also need to remain visible so that you are seen as a valued member of the group. Cinda-Sue Davis's advice, on the following page, rings true for women in any sector.

Family Issues.   A 1995 study by Sonnert and Holton of women and men in science found that both genders placed equal importance on career and family, but that women reported more tension among their various roles. Furthermore, many more women (21%) cited family demands as career obstacles than did men (3%). The authors discovered that women scientists are more often married to scientists: 62% of the women in their sample had a spouse with a Ph.D., whereas only 19% of men had a spouse with a Ph.D. (Sonnert and Holton, 1995).

There are both advantages and disadvantages to having a spouse or significant other who is equally educated and in the same employment sector. If you are both in academia, you are more likely to have similar lifestyles, but you are also likely to find it hard to find good jobs for both of you at the same university or even in the same town (the two-body problem).

ON BEING SEEN

"It's very important that people know and understand what you're doing, what you are contributing. I truly felt, when I was much younger, that if you did something well you would be noticed and appreciated, but that's not necessarily the case. You have to do a level of self-promotion that I as a woman was very uncomfortable doing—and perhaps still am uncomfortable doing. Of course, there's a fine line between coming across like you're bragging about yourself and keeping yourself visible to people around you who are influential.

"It's also important to let your peers know what you're doing, so they'll understand what your contribution is. I can remember a story about a young woman coming up for tenure and being told by somebody in the department that she would have a rough time because she obviously hadn't published very much. And she said, 'What do you mean? I've published a lot.' Apparently there was an unwritten policy in the department that when you're being published you put a copy of the paper in everybody's mail box, and she didn't know that! So the peers in her department who were not directly involved in the day-to-day research of her field really thought that she wasn't producing very much, and so they had dismissed her. Another problem here was that nobody had told her about this policy.

"It's also important to realize that your male colleagues can be just as important allies and mentors as the women can be. Oftentimes women, as a distinct minority, will seek out other women for advice and strategies. But there are very many sympathetic men out there."

Cinda-Sue Davis
*Director of the Women in Science and Engineering Program,
University of Michigan
(Ph.D., Biochemistry, University of Michigan)*

And then there are the kids. For those who want to have children, timing is a tough issue. Here you are just starting your career, but now is the prime time to start a family as well. There is no one good time to have children: some women find that having their children in graduate school is a good choice, most wait until they are in the workforce, and some in academia even wait until after they have tenure. It's a completely personal decision, no matter how hard it might be to make.

What's become easier, however, are policies in just about every workplace that allow a woman to have a decent maternity leave and not have to worry about losing her job or her place on the career ladder.

A newly tenured engineering professor had her first child before she received tenure and just recently had a second. She had a year added to her tenure clock after her first child but chose not to apply this extra year in the promotion and tenure process. She is taking a maternity leave of one term after the birth of her second

Sarah Rajala

BABY TIME

"In 1986, I was the first female faculty member to ever have a child in the College of Engineering, so there were no policies. N.C. State didn't have a maternity leave for faculty members. When I found out I was pregnant, I asked the
dean, 'Well, what are we going to do?' The dean's comment at that time was,
'Based on previous experience, let's think about it like a heart attack.' It was
kind of an interesting analogy, but he was pulling out from his mind what happens if he had a male faculty member who couldn't be in class for some medical condition. This was the approach that was used in the College of
Engineering until a university-wide policy was put into play. The policy then
gave department heads and deans common guidelines to work with the
women faculty to accommodate the need to be away from the office without
penalizing the individual.

"When should a woman faculty member have children? I don't think
there's any one right answer to that. I waited until I had tenure first—for a variety of reasons. But I didn't intentionally do that; I wasn't sure when I got out of
graduate school whether I'd even have children. But I did end up having two
children. And for me, that worked out well, but it slowed down my progress
during the time I was an associate professor. I couldn't find enough hours in
the day to keep everything going at that same hectic pace I did as an assistant
professor. But after those first couple of years when my children were very
small and things sort of settled into a pattern, I figured out how to juggle more
things. My approach is if there's something I can get someone to help me with
and I can afford to pay them to do it then I'm going to try to do that. It makes
life a lot easier! This works well for me, finding the right support system so you
can interact with your children as needed, as well as when you want to. Of
course, you have to figure out how to match your family needs with what
you're doing in your career at that moment. This approach worked for me, but I
don't think every woman should have to wait until after she's tenured to have a
child.

"More and more women faculty are having babies before they're tenured.
I think one of the things that helps—which we have here at N.C. State—is a
policy where they can stop the tenure clock for a year if they have a child before they're tenured. Some women do have concerns about that, saying,

'Some people interpret that as my having an extra year to build publications.' So, there are some potential pitfalls, but I think if institutions educate faculty and department heads, it can be a positive thing and will give women more flexibility. I'm not sure how men feel about using such a policy. I do believe the policy should be available to both men and women; for example, if a man was adopting or assuming new baby care. It shouldn't be just a female option."

Sarah Rajala
Associate Dean for Academic Affairs, College of Engineering,
North Carolina State University
(Ph.D., Electrical Engineering, Rice University)

child. Though she was the first female professor in her department to have a child, there are now several new hires (both male and female) who have small children. She is happy to report, "Some of us bring our children with us to meetings, etc. and this has been tolerated well by our colleagues university-wide."

A TIP FOR MOTHERS IN INDUSTRY

It is often more difficult in industry to reduce hours while keeping full-time benefits. If you wish to maintain your full-time status after your maternity leave is over, but wish to spend more time with your baby, put together a plan that will show your company you are still dedicated to your job but will allow you to work more flexible hours. Consider working some hours at home or having a four-day workweek with longer hours so that you can spend one day at home.

Handling Harassment. Even in the twenty-first century, even in your dream job, you still may face the nightmare of sexual harassment. And it may be hard to define. In her book *Lifting a Ton of Feathers: A Woman's Guide to Surviving in the Academic World,* Paula Caplan notes that the definition of sexual harassment no longer describes simply a demand for sexual contact, but also includes gender-based harassment such as the presence of a hostile working environment (Caplan, 1994).

An engineer who has worked in both academia and industry knows that harassment still occurs in the workplace. She notes that it is often unintentional but may be serious enough that you feel it affects your work. Depending on the situation, women choose to handle it in different ways. Informal discussions about these problems with co-workers whom you trust (female or male) can be very useful. In more serious situations, some women choose more formal routes that require documentation of events, while others decide it is not worth the energy or controversy and just move on to a new assignment or new job. "There is no one ideal way to handle harassment," she concludes, "just handle it in a way that you are comfortable with and best suits your needs."

In Academia

In an essay written for new faculty at the University of Southern California, computer science Professor George Bekey touts the joys of academia:

> I believe that the academic career is the best of all possible careers. It provides the highest ratio of satisfaction to frustration of all careers I know. It provides a life-long opportunity for learning and creativity, and for contact with stimulating colleagues and students. Each new semester is a fresh start, which renews hope and opportunity. But most of all, a university career is a continuing learning experience, always on the growing edge of new knowledge. However, it has initiation rites, and it involves hard work and a level of commitment not required in many other jobs. But the rewards are worth the effort, and I hope that you will accept the challenge and make this a long-term career. I wish you great success.

In 1995, about 27% of women graduating with doctorates in science or engineering had definite plans to enter academia. Also in 1995, as a whole, about 18% of all faculty in science and engineering were women, with only 10% of the full professors in those fields being women. The percentage of women faculty goes up when you look at the younger end of the array, with women making up 28% of the total engineering and science faculty younger than age 35 (National Science Foundation, 1999).

At the 1998 Women in Research Universities Conference, Marianne Ferber noted that while representation of women has been growing, "progress has not been continuous or steady." She attributes the lack of progress to such factors as the lack of successful female role models, the unequal distribution of women by field, and—most seriously—"the tendency to devalue women scholars in their work."

In particular, Ferber notes that men tend to view women and men scholars differently, judging comparable publications or job applications differently depending on whether the names attached were male or female (as cited in Etaugh and Kasley, 1981; Fidell, 1970; Wenneras and Wold, 1997). Male students also tended to rate male instructors more favorably than female ones, while the opposite was true of female students (Ferber and Huber, 1975; Kaschak, 1979). Ferber observes that more recent studies show that such bias remains (Ferber, 1998).

One recently tenured electrical engineering professor has noticed a marked difference between how students perceive and treat female and male professors. She agrees that the students critique the professors' teaching differently. In addition, she notes that the students call her by her first name or "Mrs." instead of "Dr." or "Professor," which she doesn't think they do to their male professors. In her research, however, she has found both undergraduate and graduate students to be productive and respectful.

In Industry

Although women make up about half of the total workforce, they make up only about 12% of the scientific and engineering workforce in industry (*Statistical Ab-*

stract of the United States, 1996). But according to a 1999 survey by Catalyst (a nonprofit organization that promotes women's involvement in industry), women are making steady (if slow) progress in Fortune 500 companies' headquarters. The survey reported that in 1999 11.9% of the corporate officers in the nation's top 500 companies were women (up from 8.7% in 1995), but that women held only 6.8% of the "line officer" positions leading to promotions.

In 1996, Catalyst also reported that 82% of CEOs cite women's lack of general management experience as holding them back from top management positions. But women executives nearly equally cite lack of management experience (47%), along with "male stereotyping and preconceptions of women" (52%) and being excluded

Tresa Pollock (Photo credit: Ken Andreyo)

ON BEING A WOMAN IN INDUSTRY

"I worked at General Electric Aircraft Engines right out of my Ph.D. program. I was one of about 20 women in a materials lab of about 600 people. Even though the numbers of women in engineering programs at universities are increasing, in the workplace you are likely to still be in the minority. Being in the minority can be an advantage or a disadvantage, depending on how you look at it. You are very visible, but you can also be drafted to serve on too many committees, internally and externally. Learn to say no, and think carefully about how you want to spend your time."

"I had a very good experience at G.E., because I got to do what I wanted technically and I learned a great deal from my colleagues there. To work in an industry environment you must be very assertive in many situations and take credit for your accomplishments. However, it is also very important to give credit to others in a team situation, particularly since major successes usually involve the contributions of many co-workers. If you are always aggressive, people will ignore you; decide which battles you really want to win."

Tresa Pollock
Professor, Materials Science and Engineering, University of Michigan
(Ph.D., Materials Science and Engineering,
Massachusetts Institute of Technology)

from informal communication networks (49%), as holding them back from those positions (Catalyst, 1996).

Again, numbers can't tell the whole story: each individual's experience will be different. After all, you will bring your own unique set of strengths to your job, and company environments differ dramatically.

Many top managers in industry have begun to realize that a diverse workforce can mean profits for the company. At the 1998 International Engineering Foundation Conference, Steve Hadden, a vice president at Texaco, explained, "Diversity and inclusiveness pay. They bring a perspective that more closely represents the markets in which we must compete. They bring more new and innovative ideas to

---

A DIVERSITY CONUNDRUM

"I work in an organization where we have a significant number of women with doctorate degrees—women physicists, women engineers, women biologists, women immunologists, women microbiologists. I find that 90% of the time the people I work with really do value the diversity of thought that the women bring. I don't find that on a day-to-day working level there are the kinds of problems that you typically hear about. I've always felt a part of a kind of intellectual core team—that we're working on the same topics, we're challenged by the same kinds of questions, and it's just fun to sit there and talk about them and think about them and do our research.

"There was the generation before me that I think really had to push hard on opening the doors and leveling the playing field and in dealing with all of the 'good old boy' networks, that kind of culture. Because of the time frame that I've entered the workforce, I think that many of the 'women in the workforce' issues have been addressed, but not eliminated. I also believe that we as women need to recognize the progress that has been made. I believe we sometimes get confused as to why there aren't a lot of females in the very senior level of management. I think that I get confused and I think my peers get confused between 'Is this a female issue?' or 'Is this an experience-based issue?' From my own personal perspective regarding my promotions, I truly believe it is an experience-based issue.

"As a female and as a female scientist I have probably been provided opportunities because there is such a strong sensitivity about diversity, at least with the company and institutes that I have worked with. I think in many ways I receive more management attention than some of my male peers; however, I've managed to maintain excellent relationships with my male peers as we have all grown in our research positions."

Andrea
manager, national lab
(Ph.D., chemical engineering)

our companies and reduce the cycle time in effectively bringing them to the market-place" (Hadden, 1999).

This viewpoint is substantiated in some recent studies. A five-year study by Lawrence A. Pfaff and Associates of 2,482 managers revealed that women managers best men in such areas as planning, facilitating change, decisiveness, teamwork, and empowering workers. Women managers also mean better stock performance, reports Theresa Welbourne, assistant professor of human resource studies in Cornell University's School of Industrial and Labor Relations. Her study showed that the stocks of companies that went public that had more women on their top management teams performed better in both the short and long term than those with few or no women at the top (Welbourne, 1999).

In the Public Sector

When he took office in the early 1990s, President Clinton promised to build a government "that looks like America." He kept his promise by making several visible appointments of women to top positions. But, as Daryl Chubin argued at a National Academy of Sciences conference, the civil service ranks are still filled with a "historical preponderance of men in 'professional' and 'administrative'" positions, with most women in the "clerical" and "technical" positions (Chubin, 1999).

Numbers do seem to be looking up, however. The most recent statistics from the U.S. Office of Personnel Management show that 40.9% of women working in the federal government hold administrative or professional jobs, including 31.3% of all jobs in the top-notch General Schedule grades 12 through 15.

WHAT THE FUTURE HOLDS FOR WOMEN SCIENTISTS AND ENGINEERS

- On the Shoulders of Giants
- Mentoring Others
- You Have Found Your Place

On the Shoulders of Giants

So you are where you want to be. You have your Ph.D. and perhaps your dream job. Perhaps you've had smooth sailing through graduate school and the job search (even the interviewing) and are now comfortable in your new position. Or perhaps you didn't have it so easy—you had to shake off the stereotypes that some people had of engineers or scientists being male, or you had to fight to get equitable lab time or field experience.

Things may have been rough, but just think of the women who came before you. At the end of the nineteenth century, even the women who completed the required graduate work were denied the degree. And even once doctoral programs in engineering and science were opened to women, other barriers remained. To coin a phrase, you stand on the shoulders of giants—and that makes it a bit easier for you to reach for the sky.

In 1998, women earned 34.3% of science and engineering doctorates (National Science Foundation, 2000). Although this figure means that men earned the other 65.7% of science and engineering doctorates that year, it shows that women are definitely making headway in entering fields that have traditionally excluded them.

Until the number of women in engineering and science reaches a critical mass—and perhaps even after—some people will maintain that women just don't belong, an attitude that can complicate your work in what are already demanding fields. But you have reached your goal: you have worked hard, received your Ph.D., and are—hopefully—doing what you've always dreamed of.

Even though your career has just begun (and there's more hard work ahead of you), just think about how far you've come: You are a smart woman with a doctorate. You are an expert in your chosen field. You are qualified to work in the highest echelons of academia, industry, or the public sector (or even do consulting). With a Ph.D., your status allows you the latitude to create new knowledge, and you will enjoy a very good salary, job security, and mobility. But, mostly, you should relish the fact that you are where you want to be, doing what you love.

Mentoring Others

Whatever your educational experience has been, don't you think it would have been easier with more support or role models than you had as you were going through graduate school, or even as an undergraduate or high school student? Now is your chance to give something back if you did receive that support, or to make a difference if you did not. You can strive to make things equitable for young women who have the same dream you did: to become an engineer or a scientist.

Cynthia Friend, the first tenured female professor of chemistry at Harvard, gives this advice: "One thing I would convey to younger women is an appreciation for how much has changed in so little time because women pushed the system, put pressure on it to make things change. Women today have to keep pushing for additional change and be vigilant about not letting things revert back" (Ambrose et al., 1997).

You can help pave the way for young women, just as the earlier women pioneers opened paths in your field for you and your contemporaries. You can mentor a budding young scientist or engineer. Or you can simply make yourself visible to show that there are women out there who are comfortable and successful in a lab coat or a hard hat.

On a local level, those of you in industry, academia, or the public sector have many opportunities to speak to groups in local junior high or high schools. In addition, many universities hold regular, informal meetings for faculty women to meet and interact with women students in engineering and science fields. You can also

Anita Borg

How to Be a Mentor and Other Advice

Anita Borg, founder of the electronic networking group Systers, computer scientist at Xerox, and president of the Institute for Women and Technology (Ph.D., Computer Science, New York University), talks about mentoring and balancing the personal and professional aspects of your life.

Why is mentoring important?

My general philosophy is that we should be mentoring other women for our entire lives. In fact, at every point we're at, we both need to be mentored and need to be mentors because that is the human way of sharing your experience and culture and knowledge. If we're interested in sharing our experience and making our experience more valuable than it has been just for us, then by sharing it in some fashion with others it's making our own experience more valuable.

But it's foolish to assume that if one mentoring relationship didn't work then mentoring is bad. It's just that the match was bad.

Is there a good balance between helping others and helping yourself?

In order to effectively help others you have to make sure that you don't damage yourself.

In terms of committees, I would say that it might be useful to look at what the average number of committees a faculty member at your level usually gets involved in. And then just put a hard limit on something that's not very much greater than that. In terms of activities for a nontenured faculty member, those committees should be things that have a pretty strong likelihood of having a positive impact on your tenured department members. So there are limits.

I think it's good to give back some, but you don't have to feel guilty if you can't be on every committee that asks you, particularly university committees. If departments are only willing to hire one woman in the department, but want a woman on every committee, then they'd better damn well hire some more women!

How can one be both a woman and an engineer or a scientist?

I think that one of the things that you have to remember is that all of the elements of a woman's experience are valuable to her career. There's not always an understanding of that within the career market. As a result, women start to feel that whatever their experience is as a whole person, only little bits of it are appropriately valuable to her career. And I really think that the most creative person brings her culture, her socialization, her intuition, her experience, her emotions in full force into the creative process, because whole people are simply more creative.

All of the work that has been done and popularized lately about emotional intelligence points to the fact that people's ability to think intelligently is enhanced when they bring the fullness of their character to the process. You will be smarter and more successful when you bring your whole self to your work. It results in a kind of self-confidence.

And the other piece of advice is to hang out with other professionals and women in one's own field and talk about all aspects of life—personal and work-oriented.

become involved at a national level by participating in activities sponsored by some of the organizations listed in the Web Resources at the back of the book.

You Have Found Your Place

Congratulations on your choice to pursue a Ph.D. in science or engineering. Your love for the work and your dedication to your field will take you far. Though sometimes it may look tough for a woman in science and engineering, always remember: if you are doing what you love, this is your place, too.

BIBLIOGRAPHY
AND REFERENCES

Adams, Howard G. *Making the Grade in Graduate School: Survival Strategy 101.* Notre Dame, IN: National Center for Graduate Education for Minorities, 1993.

Adelman, Clifford. *Women at Thirtysomething: Paradoxes of Attainment.* Washington, DC: U.S. Department of Education, Office of Educational Research and Development, 1991.

Aisenberg, N., and Harrington, M. *Women of Academe.* Amherst: University of Massachusetts Press, 1988.

Aldrich, Michele I. "Review Essay: Women in Science." *Signs: Journal of Women in Culture and Society* 4, no. 1 (1978): 126–135.

Ambrose, Susan, and Davidson, Cliff. *The New Professor's Handbook: A Guide to Teaching and Research in Engineering and Science.* Bolton, MA: Anker Publishing Co., 1994.

Ambrose, S. A., Dunkle, K. L., Lazarus, B. B., Nair, I., and Harkus, D. *Journeys of Women in Science and Engineering: No Universal Constants.* Philadelphia: Temple University Press, 1997.

Anderson, M. S. "Toward a Profile of the Highly Collaborative Graduate Programs. Effects on the Doctoral Experience and Departmental Environment." Paper presented at the annual meeting of the Association for the Study of Higher Education, Pittsburgh, PA, 1993.

Barber, L. A. "U.S. Women in Science and Engineering, 1960–1990: Progress Toward Equity?" *Journal of Higher Education* 66, no. 2 (March–April 1995): 213–234.

Berg, H. M., and Ferber, M. A. "Men and Women Graduate Students: Who Succeeds and Why?" *Journal of Higher Education,* 54 (1983): 629–648.

Boyer, Ernest L. *Scholarship Reconsidered: Prioritizing the Professoriate.* The Carnegie Foundation for Advancement of Teaching, 1990.

Broome, Taft. "Heroic Woman Engineer." In Anne M. Humphreys (Ed.), *Bridging the Gender Gap in Engineering and Science: Conference Proceedings.* Pittsburgh: New Image Press, 1996.

Burke, R. J. "Mentors in Organizations." *Group & Organization Studies* 9, no. 3 (September 1984): 353–372.

Burke, R. J. "Relationships, In and Around Organizations: It's Both Who You Know and What You Know That Counts." *Psychological Reports* 55 (1984): 299–307.

89

Caplan, Paula J. *Lifting a Ton of Feathers: A Woman's Guide to Surviving in the Academic World.* Toronto, Canada: Council of Ontario Universities, 1994.

Catalyst. "Women in Corporate Leadership: Progress and Prospects." Survey Report, 1996.

Chamberlain, M. K. *Women in Academe: Progress and Prospects.* New York: Russell Sage Foundation, 1988.

Charles, B. L. "Institutional Environment: A Critical Factor for Minorities in the Graduate Education Pipeline." In J. M. Jones, M. E. Goertz, and C. V. Kuh (Eds.), *Minorities in Graduate Education: Pipeline, Policy and Practice* (pp. 70–75). Princeton, NJ: Educational Testing Service, 1992.

Chubin, Daryl E. "What Policies of Government Agencies Are Worth Emulating?" In Cecily Cannan Selby (Ed.), *Women in Science and Engineering: Choices for Success. Annals of the New York Academy of Sciences* 869 (1999): 197–199.

Clark, S. M., and Corcoran, M. "Perspectives on the Professional Socialization of Women Faculty: A Case of Accumulative Disadvantages?" *Journal of Higher Education* 57, no. 1 (1986): 20–43.

Clewell, Beatriz C., and Anderson, Bernice. *Women of Color in Mathematics, Science, and Engineering: A Review of the Literature.* Washington, DC: Center for Women Policy Studies, 1991.

Council of Graduate Schools. *Enhancing the Minority Presence in Graduate Education: Models and Resources for Minority Student Recruitment and Retention* (Vol. 4). Washington, DC: Council of Graduate Schools, 1992.

Daniels, Jane Zimmer. "Population and Pipeline." In Anne M. Humphreys (Ed.), *Bridging the Gender Gap in Engineering and Science: Conference Proceedings.* Pittsburgh: New Image Press, 1996.

Davis, C. S., Ginorio, A. B., Hollenshead, C. S., Lazarus, B. B., Rayman, P. M., and Associates. *The Equity Equation: Fostering the Advancement of Women in the Sciences, Mathematics and Engineering.* San Francisco: Jossey-Bass Publishers, 1996.

Etaugh, Claire, and Kasley, Helen C. "Evaluating Competence: Effects of Sex, Marital Status and Parental Status." *Psychology of Women Quarterly* 6, no. 2 (1981): 196–203.

Etzkowitz, H., Kemelgor, C., Neuschatz, M., Uzzi, B., and Alonzo, J. "Athena Unbound: Barriers to Women in Academic Science and Engineering." *Science and Public Policy* 19, no. 3 (1992): 157–179.

Etzkowitz, H., Kemelgor, C., Neuschatz, M., Uzzi, B., and Alonzo, J. "The Paradox of 'Critical Mass' for Women in Science." *Science* 266 (1994): 51–54.

Fausto-Sterling, A. *Myths of Gender: Biological Theories about Women and Men.* Rev. ed. New York: Basic Books, 1992.

Feldman, Saul. *Escape from the Doll's House: Women in Graduate and Professional Education.* New York: McGraw-Hill, 1974.

Ferber, Marianne. "Women's Uneven Progress in Academia: Problems and Solutions." Presentation, Women in Research Universities Conference, Cambridge, MA, 1998.

Ferber, Marianne A., and Huber, Joan A. "Sex of Student and Instructor: A Study of Student Bias." *American Journal of Sociology* 80, no. 4 (January 1975): 949–963.

Fidell, L. S. "Empirical Verification of Sex Discrimination in Hiring Practices in Psychology." *American Psychologist* 25, no. 12 (December 1970): 1094–1098.

Fort, D.C. (Ed.). *A Hand Up: Women Mentoring Women in Science*. Washington, DC: Association for Women in Science, 1993.

Fox, Mary Frank. "Women and Higher Education: Gender Differences in the Status of Students and Scholars." In J. Freeman (Ed.), *Women: A Feminist Perspective*, 5th ed. (pp. 220–237). Mountain View, CA: Mayfield, 1995.

Fox, Mary Frank. "Gender, Faculty and Doctoral Education: Departmental Cultures and Climate." Presentation, Women in Research Universities Conference, Cambridge, MA, 1998.

Frank, M. J. Improving the Climate for Women in Physics Departments. Unpublished proposal for a national study, 1994.

Frieze, I. H. "Psychological Barriers for Women in Sciences: Internal and External." In J. A. Ramaley (Ed.), *Covert Discrimination and Women in the Sciences* (pp. 65–95). Boulder, CO: Westview Press, 1978.

Geppert, L. "The Uphill Struggle: No Rose Garden for Women in Engineering." *Association for Women in Science* 25, no. 2 (March/April 1996): 16–17.

Glassick, Charles E., Huber, Mary Taylor, and Maeroff, Gene I. *Evaluation of the Professoriate*. San Francisco: Jossey-Bass Publishers, 1997.

Glazer, Penina Migdal, and Slater, Miriam. *Unequal Colleagues: The Entrance of Women into the Professions, 1890–1940*. New Brunswick, NJ: Rutgers University Press, 1987.

Grayson, Lawrence P. "A Brief History of Engineering Education in the United States." *Engineering Education* 68 (December 1977): 246–268.

Green, Judy, and LaDuke, Jeanne. "Women in the American Mathematics Community: The Pro 1940 Ph.D.'s." *The Mathematical Intelligencer* 9, no. 1 (1987).

Hadden, Steve, "The Profitable Workplace of the Twenty-first Century." In Barbara Bogue, Priscilla Guthrie, Barbara Lazarus, and Steve Hadden (Eds.), *Tackling the Engineering Resources Shortage: Creating New Paradigms for Developing and Retaining Women Engineers*. SPIE (International Society for Optical Engineering) 44 (1999): 4–5.

Harding, S. L. *Whose Science? Whose Knowledge?: Thinking from Women's Lives*. Ithaca, NY: Cornell University Press, 1991.

Harding, Sandra, and O'Barr, Jean F. (Eds.). *Sex and Scientific Inquiry*. Chicago: University of Chicago Press, 1987.

Hollenshead, Carol. "Women in the Academy: Confronting Barriers to Equality." Presentation, Women in Research Universities Conference, Cambridge, MA, 1998.

Hornig, L. S. "Women in Science and Engineering: Why So Few?" *Technology Review* 87, (1984): 30–41.

Hornig, L. S. "Women Graduate Students: A Literature Review and Synthesis." In L. S. Dix (Ed.), *Women: Their Under-representation and Career Differentials in Science and Engineering* (pp. 103–122). Washington, DC: National Academy Press, 1987.

James, Edward T., James, Janet Wilson, and Boyer, Paul S. (Eds.). *Notable American Women*. Cambridge, MA: Harvard University Press, 1971.

Jerrard, Richard, and Jerrard, Margot. *The Grad School Handbook: An Insider's Guide to Getting In and Succeeding*. New York: Perigee, 1998.

Kantrowitz, M., and Digennaro, J. P. *Prentice Hall Guide to Scholarships and Fellowships for Math and Science Students: A Resource for Students Pursuing Careers in Mathematics and Science*. Englewood Cliffs, NJ: Prentice Hall, 1993.

Kaschak, Ellyn. "Sex Bias in Student Evaluations of College Professors." *Psychology of Women Quarterly* 2, no. 3 (1979): 236–244.

Kass-Simon, Gabrielle, Farnes, Patricia, and Nash, Deborah (Eds.). *Women of Science: Righting the Record.* Bloomington: Indiana University Press, 1990.

Keller, Evelyn Fox. *Reflections on Science and Gender.* New Haven, CT: Yale University Press, 1985.

Kuh, Charlotte V. "You've Come a Long Way: Data on Women Doctoral Scientists and Engineers in Research Universities." Presentation, Women in Research Universities Conference, Cambridge, MA, 1998.

Lomperis, A. M. T. "Are Women Changing the Nature of the Academic Profession?" *Journal of Higher Education* 61 (November/December 1990): 643–677.

Malcom, S. M. "Increasing the Participation of Black Women in Science and Technology." *SAGE* 6, no. 2 (1989): 15–17.

Malcom, S. M., Hall P. Q., and Brown, J. W. *The Double Bind: The Price of Being a Minority Woman in Science* (AAAS Report No. 76-R-3). Washington, DC: American Association for the Advancement of Science, 1976.

Mandula, Barbara. "Women Scientists Still Behind." *Association for Women in Science Magazine* 20 (May–June 1991): 10–11.

Mapstone, E. R. "Rational Men and Disagreeable Women—The Social Construction of Argument: Women Argue Differently from Men." Paper given at the British Psychological Society Conference, City University, London (as cited in Phillips and Pugh in *How to Get a Ph.D.*), 1990

Matyas, M. L., and Malcom, S. M. (Eds.) *Investing in Human Potential: Science and Engineering at the Crossroads.* Washington, DC: American Association for the Advancement of Science, 1991.

MIT Computer Science Female Graduate Students and Research Staff. *Barriers to Equality in Academia: Women in Computer Science at MIT.* February 1983. (Copies available from EECS Graduate Office at MIT, Cambridge, MA 02139)

National Science Foundation. "Women, Minorities, and Persons with Disabilities in Science and Engineering: 1998." (NSF 99-338). Arlington, VA, 1999.

National Science Foundation, Division of Science Resources Studies. "Science and Engineering Doctorate Awards: 1998." (NSF 00-304). Arlington, VA, 2000.

Nyquist, J. D., Abbott, R. D., Wulff, D. H., and Sprague, J. *Preparing the Professoriate of Tomorrow to Teach: Selected Readings in TA Training.* Dubuque, IA: Kendall/Hunt, 1991.

Pearson, C. S., Shaflik, D. L., and Touchton, J. G. (1989). *Educating the Majority: Women Challenge Tradition in Higher Education.* New York: Macmillan, 1989.

Phillips, E. M., and Pugh, D. S. *How to Get a Ph.D.: A Handbook for Students and Their Supervisors.* Buckingham, England: Open University Press, 1994.

Rhode, D. L. *Speaking of Sex: The Denial of Gender Inequality.* Cambridge, MA: Harvard University Press, 1997.

Rich, Adrienne. "Toward a Woman-Centered University: 1973–74." In *On Lies, Secrets and Silence: Selected Prose (1966–1978).* New York: W. W. Norton, 1979.

Richards, Robert H. *Robert Hallowell Richards: His Mark.* Boston: Little, Brown, 1936.

Rosser, S. V. *Female-Friendly Science: Applying Women's Studies Methods and Theories to Attract Students.* New York: Pergamon Press, 1990.

Rossi, Alice. "Women in Science: Why So Few?" *Science* 148, no. 3674 (1965).

Rossi, Alice. "Report of Committee W 1970-71." *AAUP Bulletin,* Summer 1971.

Rossi, Alice. "Seasons of a Woman's Life." In Bennett M. Berger (Ed.), *Authors of Their Own Lives: Intellectual Autobiographies by Twenty American Sociologists.* Berkeley: University of California Press, 1990.

Rossiter, Margaret W. *Women Scientists in America: Struggles and Strategies to 1940.* Baltimore, MD: Johns Hopkins University Press, 1982.

Rossiter, Margaret W. *Women Scientists in America: Before Affirmative Action 1940–1972.* Baltimore, MD: Johns Hopkins University Press, 1995.

Rowe, Mary. "The Saturn's Rings Phenomenon: Micro-Inequities and Unequal Opportunity in the American Economy." *Proceedings of NSF Conference on Women's Leadership and Authority.* Santa Cruz: University of California–Santa Cruz, 1977.

Sadker, Myra, and Sadker, David. *Failing at Fairness: How America's Schools Cheat Girls.* New York: Charles Scribner's Sons, 1994.

Sanderson, A., Dugoni, B., Hoffer, T., and Selfa L. *Doctorate Recipients from United States Universities: Summary Report 1998.* Chicago: National Opinion Research Center. (The report gives the results of data collected in the Survey of Earned Doctorates, conducted for five federal agencies: NSF, NIH, NEH, USED, and USDA, by NORC.) 1999.

Schiebinger, Londa. *Has Feminism Changed Science?* Cambridge, MA. Harvard University Press, 1999.

Selby, Cecily Cannan (Ed.). *Women in Science and Engineering: Choices for Success. Annals of the New York Academy of Sciences* 869, 1999.

Seymour, E. "Revisiting the 'Problem Iceberg': Science, Mathematics, and Engineering Students Still Chilled Out." *Journal of College Science Teaching* (May 1995): 392–400.

Skidmore, L. "CWSE Findings on Female Scientists and Engineering Employed in Industry." *Association for Women in Science* 25, no. 2 (March–April 1996): 6–7.

Sonnert, G., and Holton, G. *Who Succeeds in Science? The Gender Dimension.* New Brunswick, NJ: Rutgers University Press, 1995.

Sonnert, G., and Holton, G. "Career Patterns of Women and Men in the Sciences." *American Scientist* 84 (January–February 1996): 63–71.

Statistical Abstract of the United States. Washington, DC: U.S. Government Printing Office, 1996.

Traweek, Sharon. *Beamtimes and Lifetimes: The World of High Energy Physicists.* Cambridge, MA: Harvard University Press, 1992.

Trescott, M. M. "Women in the Intellectual Development of Engineering." In G. Kass-Simon & P. Farnes (Eds.), *Women of Science: Righting the Record* (pp. 59–74). Ann Arbor: University of Michigan Press, 1990.

Ulku-Steiner, B. S. "Women and Men in Doctoral Education: Students' Experiences in Gender-Balanced and Male-Dominated Programs." Ph.D. dissertation, University of North Carolina at Chapel Hill, 1997.

Vetter, B. M. "Changing Patterns of Recruitment and Employment." In V. B. Haas and C. C. Perrucci (Eds.), *Women in Scientific and Engineering Professions* (pp. 59–74). Ann Arbor: University of Michigan Press, 1984.

Vetter, B. M. (Ed.). *Professional Women and Minorities: A Total Human Resource Data Compendium.* 11th ed. Washington, DC: Commission on Professionals in Science and Technology, 1994.

Welbourne, Teresa. "Wall Street Likes Its Women: An Examination of Women in the Top Management Teams of Initial Public Offerings." Cornell University's ILR Working Paper Series, 1999.

Wenneras, Christine, and Wold, Agnes. "Nepotism and Sexism in Peer-Review." *Nature* 387 (May 1997): 341–343.

Widnall, Sheila E. "AAAS Presidential Lecture: Voices from the Pipeline." *Science* 241 (September 30, 1988): 1740–1745.

Winston, Patrick H. "Breaking Writer's Cramp." Unpublished work.

Zuckerman, Harriet, and Cole, Jonathan. "Women in American Science." *Minerva* 13, no. 1 (Spring 1975): 82–102.

Zuckerman, Harriet, Cole, Jonathan R., and Breur, John T. (Eds.). *The Outer Circle: Women in the Scientific Community.* New York: W. W. Norton, 1991.

Zwick, R. *Differences in Graduate School Attainment Patterns Across Academic Programs and Demographic Groups* (Research Rep. No. 143). Princeton, NJ: Educational Testing Service, 1991.

WEB RESOURCES

ORGANIZATIONS

Association for Women in Computing

Includes the humorous and helpful column "Live Wire: Computer Confidence for Women," informative links, and other items.

http://www.awc-hq.org/

Association for Women in Science

Dedicated to achieving equity and full participation for women in science, technology, engineering, and math.

http://www.awis.org/

Catalyst

Works with business to advance women.

http://www.catalystwomen.org/home.html

The Institute for Women and Technology

Works to advance technology that will have a positive impact on women worldwide.

http://www.iwt.org/

Institute of Electrical and Electronics Engineers (IEEE)

Promotes the creation, development, integration, sharing, and application of knowledge about electrical and information technologies and sciences for the benefit of humanity and the profession.

http://www.ieee.org

MentorNet

MentorNet (the National Electronic Industrial Mentoring Network for Women in Engineering and Science) was established in 1998 to pair undergraduate and graduate women students to women working in industry.

http://www.mentornet.net

NAFSA: Association of International Educators

Promotes the exchange of students and scholars to and from the United States. While their website provides information mostly for teachers and advisors of international students, it also provides useful links for the students themselves.

http://www.nafsa.org/

National Science Foundation

Provides information on the foundation, lists employment opportunities, and more.

http://www.nsf.gov/

National Society of Black Engineers (NSBE)

Established to "increase the number of culturally responsible black engineers who excel academically, succeed professionally and positively impact the community." Their site lists continuing education programs and conferences and will post your resume.

http://www.nsbe.org/

Office of Scientific and Engineering Personnel of the National Research Council

Analyzes the human resources needed to support science and engineering in the United States, with particular attention to graduate education and employment. It also administers fellowship and research associateship programs for federal agencies and private foundations.

http://www4.nas.edu/osep/osephome.nsf

Society for Women Engineers (SWE)

An organization for women engineers worldwide. Their site contains a variety of resources and information and even allows job seekers to post their resumes on line.

http://www.swe.org/

Systers

An organization and on-line discussion group for women in computing.

http://www.systers.org/

Women and Computer Science

Tips and Internet resources for women in computer science.

http://www.ai.mit.edu/people/ellens/gender.html

Women In Engineering Program Advocates Network (WEPAN)

Founded in 1990 to foster an engineering climate more conducive to women.

http://www.web.mit.edu/wepan/www/

Women in Technology International (WITI)

WITI.com, the "virtual home for women in technology," offers information ranging from job postings to biographies of women in technology.

http://www.witi.com

OTHER USEFUL SITES

FastWeb

A popular scholarship search site.

http://www.fastweb.com

FinAid

A good, general financial aid reference site.

http://www.finaid.org

Graduate Student Survival Guide

Provides excellent tips on such important skills as "getting the most out of your research advisor or boss" and provides links to other useful survival-guide sites.

http://www-smi.stanford.edu/people/pratt/smi/advice.html

Survival in the Academy

Essays and tips from various sources, compiled by graduate students at the University of Indiana, Bloomington.

http://dcslab.snu.ac.kr/~ilhwan/grad/survival/part1.9.html

Tomorrow's Professor

Both a website and a listserv that you can subscribe to. Provides articles and tips from those who are currently working in academia to those who aspire to work there.

http://sll.stanford.edu/projects/tomprof/home.html

U.S. Department of Education Financial Aid Page

Contains information and links on grants, loans, and scholarships.

http://www.ed.gov/finaid.html

Women's Home Page

A collection of papers and articles on women and science.

http://www.mit.edu:8001/people/sorokin/women

INDEX

ABOUT THE AUTHORS

Barbara Lazarus (Photo credit: Ken Andreyo)

Barbara Lazarus is the associate provost for Academic Affairs and an adjunct professor of educational anthropology at Carnegie Mellon University in Pittsburgh, Pennsylvania. She holds a doctorate in educational anthropology from the University of Massachusetts. Dr. Lazarus has been involved in service for and research on women in science and engineering since the early 1980s. For 20 years, she has also done applied research on nontraditional occupations in Asia and more recently on women in science and engineering in Asia. She has spoken widely about these issues at national and international conferences. Her most recent publications include *Journeys of Women in Science and Engineering: No Universal Constants* (with Susan Ambrose et al., Temple University Press, 1997) and *The Equity Equation: Fostering the Advancement of Women in the Sciences, Mathematics and Engineering* (with Cinda-Sue Davis et al., Jossey Bass, 1996). At Carnegie Mellon, Dr. Lazarus is responsible for establishing academic policies for undergraduate and graduate students and for supervising academic support programs in a number of areas, including intercultural communications, undergraduate research, scholarships and fellowships, and academic support programs for women and students of color. Dr. Lazarus serves as a member of the Asian Women's Studies Committee and on the Advisory Committee of MentorNet, an electronic mentoring program for college students and post-docs. She serves on the board of directors of Women in Engineering Program Advocates Network (WEPAN) and, in 2000, received the WEPAN Founders Award for outstanding contributions to women in science and engineering.

Lisa Ritter (Photo credit: Ken Andreyo)

Lisa Ritter is a communications consultant at Carnegie Mellon University and is the editor of the quarterly graduate newsletter on campus. She has worked in communications for 14 years, the past four years of which have been at Carnegie Mellon, where she has also been a public relations director and coordinator of professional development seminars for graduate students. She received a bachelor's degree in English from Virginia Tech and a master's degree in professional writing from Carnegie Mellon.

Susan Ambrose (Photo credit: Ken Andreyo)

Susan Ambrose is associate provost for Educational Development, director of the Eberly Center for Teaching Excellence, and a principal lecturer in the Department of History at Carnegie Mellon University. She received her doctorate in history from Carnegie Mellon in 1986. The Eberly Center assists faculty and graduate students in all aspects of teaching, from curriculum design to formative evaluation, by advising about cognitive and educational research. Her research interests include applying cognitive principles to education and understanding how class origin, sex, race and ethnicity, social conceptions of women, and other variables collectively influence women's life decisions and careers in engineering and science. She is lead author of *Journeys of Women in Engineering and Science: No Universal Constants* (Temple University Press, 1998) and coauthor of *The New Professor's Handbook:*

An Introductory Guide to Teaching and Research in Engineering and Science (Anker Publishing, 1994). She conducts workshops across the country and abroad on a variety of teaching and learning topics, and actively consults with colleges and universities creating or revamping faculty development programs. Dr. Ambrose has codirected Carnegie Mellon's National Science Foundation Engineering Education Scholars Workshop (1996–2000), a national program aimed at helping new engineering Ph.D.s transition into faculty life. She has also served as a visiting scholar for the American Society of Engineering Education (1998 and 2000), visiting engineering schools to share her expertise in engineering education. She was recently honored with an American Council on Education fellowship for the 1999–2000 academic year.